Intergovernmental Oceanographic Commission
technical series

AF784598

20

A Focus for Ocean Research

Intergovernmental Oceanographic Commission
History, Functions, Achievements

By Dr. Hans Ulrich ROLL

Unesco, 1979

The designations employed and the presentation of the material in this publication do not imply the expression of any opinion whatsoever on the part of the Secretariats of Unesco and IOC concerning the legal status of any country or territory, or of its authorities, or concerning the delimitations of the frontiers of any country or territory,

ISBN 92-3-101754-3

French edition 92-3-201 754-7
Spanish edition 92-3-301 754-7
Russian edition 92-3-401 754-4

Published in 1979
by the United Nations Educational,
Scientific and Cultural Organization,
7, Place de Fontenoy, 75700-Paris

Composed by Europrom and
printed in Unesco's workshops

Foreword

The preparation of this publication is the result of the timely and fortuitous combination of several factors. In the first place, after almost two decades of existence, it is an opportune moment to disseminate to the international scientific community information on the fundamental aspects, the principal programmes and the major achievements of the Commission. Another factor is the expansion of the Commission during this period. Since its inception the number of IOC Member States has increased considerably, from a total of 40 in 1961 to 103 today.

This substantial growth in membership, principally of developing countries, has produced a considerable impact on the structure and functions of IOC. As a consequence, the Commission has recently undertaken an evaluation of its activities in order to define, in a clear fashion, its role and future functions in light of both this growth and the new emerging legal regime for the oceans. The decision to conduct this self-evaluation should be interpreted as a result of the concern expressed by a large number of Member States that the Commission be strengthened. It is evident that without abandoning responsibility for promoting basic or fundamental scientific investigation on a global scale, the IOC is responding to the wider needs and interests of all Member States.

This new trend can be seen in the adoption of a regional approach to the formulation and implementation of co-operative investigations, as well as in greater emphasis on certain resource-oriented research activities. This will certainly result in a number of changes, particularly in the planning of its programmes, which will be more attuned to the interests of developing countries. Certain administrative and financial arrangements will also have to follow suit.

During the second of these two decades of the Commission's existence, the Third United Nations Conference on the Law of the Sea (UNCLOS) has – with the participation of 158 countries – been formulating a new legal regime for ocean space. In fact the deliberations at this Conference and at its predecessor – The Committee on the Peaceful Uses of the Sea-bed and the Ocean Floor beyond the limits of National Jurisdiction – are the first occasions where the conduct of marine scientific research has been addressed at a diplomatic level.

The Conference has, at the time of writing, reached consensus on the concept of a 200-mile exclusive economic zone within which marine scientific research can only be conducted with the consent of the coastal state. Furthermore, taking advantage of this consensus, a very large number of coastal states has already adopted jurisdictional maritime zones extending up to 200 miles from their coasts. The unilateral establishment of these 200-mile maritime zones has also engendered the formulation and application of legal provisions regulating the conduct of marine scientific research activities within these zones. This system which departs from the so-called freedom of scientific research approach, or the notification regime, will necessitate procedures quite different from those under which marine scientific research activities have been carried out in the past.

An important part of the new legal regime for ocean space is that competent international organizations – such as IOC – are recognised as having the right, for the first time in their history, to undertake marine scientific research activities by themselves. In the past, this right had only been recognized as belonging to the States. This is obviously of great importance for the formulation of the Commission's future functions and programmes.

As a result of these new developments, it is likely that the Intergovernmental Oceanographic Commission of the future will be very different from that of today. The degree of difference will depend a great deal upon the decisions made by the Member States of IOC based upon the final results of the Third United Nations Conference on the Law of the Sea. This current study will thus be a very valuable reference tool during the evaluation process, as it is essential to look to the past when planning for the future.

Another reason for the issue of this publication is the importance, now more than ever, of educating the world marine scientific community and the general public about the purpose of the IOC, what has been its history and what are its most important achievements. Therefore, I would like to commend and thank Dr Hans Ulrich Roll, former First Vice-Chairman of the Commission, for the timely conclusion of this study. This is a sincere and serious effort to present a clear and distinct picture of the functions and activities of the Intergovernmental Oceanographic Commission.

By analyzing its functions and programmes at this moment in history, the Commission is demonstrating its capacity to evolve and to be open to consideration of any changes which will increase its ability to serve the interests of the global marine scientific community. Indeed, marine scientific research activities are of utmost importance for the future of mankind. Such activities will play a decisive role in the solution of many of the critical problems confronting today's society in a variety of areas, such as food, energy, pollution and shelter. These activities are growing in complexity, requiring a more efficient co-ordination of information, manpower and resources. The role of co-ordination which IOC is called upon to perform takes into account the fact that marine scientific research is essentially an international undertaking.

I sincerely believe that this publication will contribute to a better understanding of the functions, programmes and goals of the Intergovernmental Oceanographic Commission.

Mexico City, 7 May 1979

Agustín Ayala-Castañares
Chairman
Intergovernmental Oceanographic Commission

Contents

CHAPTER I

The origins

1 What is IOC ?

The Intergovernmental Oceanographic Commission (IOC) is an autonomous body established within the United Nations Educational, Scientific and Cultural Organization (UNESCO) in 1960. Its purpose is «to promote scientific investigation with a view to learning more about the nature and resources of the oceans through the concerted action of its members». It further functions as *the* co-ordinating body within the United Nations system for marine science and related activities. Co-operative scientific investigations of the ocean as well as world-wide ocean services, combined with a programme of training, education and mutual assistance, form the chief components of IOC's work.

Membership of the Commission is open to any state that is a member of any of the organizations of the United Nations system. At present, January 1979, 103 countries co-operate under this arrangement (Annex I). The Secretariat of the IOC is located in Unesco Headquarters in Paris.

2 Why study the oceans ?

Mankind is beginning to recognize that the ocean is more than just a common medium for shipping, transport, trade, fishing, sea salt extraction and naval warfare which it has been for thousands of years. The fundamental role which the ocean plays in the water cycle of the earth, and as the great moderator of the earth's weather and climate regimes, is only now starting to be understood. And as the marine environment is being increasingly endangered by various forms of pollution by man, consciousness is being raised everywhere of the fact that the ocean and its sea floor represent a vast reservoir of food, mineral resources and energy, for which a rational plan of exploration and exploitation is more important now than ever.

Further, it is well known that ocean waves, currents and surges affect shipping and fishing, and may do considerable damage to the coasts of continents and islands. Adequate knowledge of these marine phenomena is of importance to owners and masters of ships, ocean engineers, naval architects and harbour designers.

These are only some of the more compelling reasons which call for a comprehensive effort in marine science. In addition, there is the sheer curiosity of unveiling the secrets of nature, which is the chief motivation of many scientists engaged in research. Studying the natural processes in the oceans and investigating marine resources are among the greatest and most fascinating tasks man is faced with at present. If he fails to recognize and to take up this «historic challenge», the human race will be unable to manage and exploit adequately and successfully the «common heritage of mankind», as the oceans are often called.

3 How to study the oceans ?

Oceanography is the scientific study of the seas and oceans. Its purpose in detail is to increase our knowledge and our understanding of all natural processes and conditions in the ocean and on and under the sea floor, namely :

- the shape, structure and history of the ocean bottom and its subsoil ;
- the stratification, motion and circulation of ocean waters under external influences and internal forces ;
- the chemical composition of ocean water including biochemical and other transport of substances ; and
- the life processes in the ocean, in particular the life cycles of organisms.

Oceanography is distinguished by the following :

- it is not a fundamental science as such, like physics or chemistry, but consists of scientific activities within which the fundamental sciences of mathematics, physics, chemistry and biology are applied ;
- it is, therefore, multidisciplinary by definition, embracing all the natural sciences in their application to the ocean ;
- it has close connexions with meteorology, the science of the earth's gaseous envelope, and with geology and geophysics which deal with the outer part of the solid earth ;
- it cannot be based on experiments under strictly controlled conditions but has to rely on measurements in nature where it is not possible to isolate the process under study from the influence of secondary factors. This explains why a vast amount of systematically collected obervational data is necessary.
- In addition, the hydrodynamic-numerical modelling of ocean processes, based on well-established physical laws, is a very important tool of investigation in oceanography. This requires the use of powerful computers and subsequently an adequate data set by which to check the results.

4 Historical roots of IOC

Oceanography is a comparatively young science. The voyages of the great discoverers up to the 18th century have greatly contributed to our knowledge of the oceans ; but this information mostly concerned the oceanic boundaries, the coasts and islands, and was confined to phenomena at the sea surface. The sea floor and the processes in the depths represented a mysterious world.

The year of birth of modern oceanography may be considered to have been 1853. At that time the ingenious and experienced US naval officer *Matthew Fontaine Maury* succeeded in persuading the participants of an

Table 1 — Major Research Cruises during the First Phase of Ocean Exploration

Ship	Country	Period	Area
Galathea	Denmark	1845-47	Circumnavigation of the globe
Challenger	UK	1872-76	" "
Gazelle	Germany	1874-76	" "
Tuscarora	USA	1874-75	Pacific Ocean
Vøringen	Norway	1876	Norwegien Sea
Blake	USA	1877-80	Western North Atlantic Ocean
Romanche	France	1883	Atlantic Ocean
Enterprise	USA	1883-86	Circumnavigation
Albatross	USA	1883-05	Atlantic +Pacific Ocean
Hirondelle Princess Alice	Monaco	1885-14	North Atlantic Ocean
Buccaneer	UK	1886	Atlantic Ocean
Vitiaz	Russia	1886-89	Circumnavigation
National	Germany	1889	Atlantic Ocean
Fram	Norway	1893-96	Arctic Ocean
Ingolf	Denmark	1895-96	North Atlantic Ocean
Belgica	Belgium	1898	South Atlantic Ocean, Antarctica
Valdivia	Germany	1898-99	Atlantic +Indian Ocean
Gauss	Germany	1901-04	Atlantic +Indian Ocean, Antarctica
Antartic	Sweden	1902-03	Atlantic, Antarctic waters
Michael Sars	Norway	1902-03	Norwegian Sea
Scotia	UK	1902-04	Atlantic, Antarctic waters
Planet	Germany	1906	Atlantic, Indian, Pacific Ocean
Pourquoi-pas?	France	1908-10	Antarctic waters
Thor	Denmark	1908-10	Mediterranean Sea
Deutschland	Germany	1911-12	Atlantic, Antarctic waters
Margarete	Denmark	1913	North Atlantic Ocean
Armauer Hansen	Norway	1913-14	" " "

Table 2 — Major Research Cruises during the Second Phase of Ocean Exploration

Ship	Country	Period	Sea Area
Dana	Denmark	1920-22	Sargasso Sea
Armauer Hansen	Norway	1922-35	Norwegian Sea
Meteor	Germany	1925-27	South Atlantic Ocean
Carnegie	USA	1928-29	Pacific Ocean
Dana	Denmark	1928-30	Circumnavigation
Willebrord Snellius	Netherlands	1929-31	South-east Asian waters
William Scoresby	UK	1929-31	Antarctic waters
Discovery	UK	1929-31	"
Meteor	Germany	1929-35	Waters around Iceland
Norwegia	Norway	1930-31	Antarctic waters
Shintoku Maru	Japan	1930-33	North Pacific Ocean
Discovery Maru	Japan	1930-33	North Pacific Ocean
Discovery II	UK	1932-33	Antarctic waters
Atlantis	USA	1932-38	Western North Atlantic Ocean, Gulf of Mexico
Meteor	Germany	1937-38	Eastern North Atlantic Ocean

international oceanographic conference in Brussels to agree upon a uniform scheme of collecting oceanic observations taken by nautical officers on shipboard on a voluntary basis. These data were processed in hydrographic and meteorological centres of the maritime countries and formed the basis for a great number of oceanic atlases, sailing directions and pilot charts which statistically describes the natural conditions at sea. This information, though, was still confined to sea surface phenomena, eg currents, temperature, ice, meteorological interactions.

Scientific investigation of the deep ocean started about 100 years ago when the British steam corvette H.M.S. *Challenger* set out to sea in December 1872 for her 3 1/2 year-long cruise round the world in an attempt to examine the ocean in its depth from the physical, chemical and biological points of view. This voyage opened the *first phase* of ocean exploration which was characterized by long, wide-ranging cruises of research ships from a great many seafaring nations, often circumnavigating the globe and collecting data and samples from vast sea areas (see Table 1). Although the measurements taken provided important single point information, they were of a haphazard or random nature and quite inadequate for the derivation of a realistic concept of, for example, oceanic circulation in the open sea areas. This phase ended with World War I.

The *second phase* of oceanography began in 1925 when the German research vessel *Meteor* undertook the first systematic investigation of a limited oceanic area, the South Atlantic Ocean, from 1925 to 1927. The result of this cruise was a general picture of the shape of the ocean floor and of the mean state and motion of the water masses in this ocean, their stratification and circulation. Similar systematic investigations were carried out subsequently in other oceanic areas (see Table 2). This second phase of ocean exploration, characterized by intensive surveys of limited oceanic areas undertaken by a single research vessel, lasted until World War II.

At the end of this period it gradually became obvious that this single-ship approach was inappropriate for obtaining a complete understanding of oceanic processes. The oceans are restless, with a great variety of motions not only at the sea surface but also at depth. This makes it unlikely that measurements taken by a single ship at different places at different times can be combined to form a realistic picture of oceanic circulation and stratification. Oceanographers learned that it is necessary to measure the representative oceanic properties simultaneously at many places if the sea area under study is to be properly investigated. Such a network of stations occupied simultaneously cannot be achieved by a single research ship but calls for a closely co-ordinated joint observational programme using a number of ships. Moreover, since the number of research vessels in any one country is limited, it became essential to pool the efforts of several nations. This was the starting point of the *third phase* of ocean exploration, characterized by international co-operation in joint programmes and which began after World War II and is ongoing. The use of research vessels has been supplemented and further developed technologically by the introduction of anchored buoy systems, drifting buoys fitted with drogues, moored sub-surface systems and observational aircraft and satellites.

A first example of such international co-operation was the investigation of the *Atlantic Polar Front* in 1958, organized by the *International Council for the Exploration of the Sea (ICES)* with headquarters in Copenhagen and carried out within the framework of the *International Geophysical Year 1957/58*. Twenty-two research vessels from eight countries participated in this co-operative investigation of the northern North Atlantic Ocean (see Table 3). Vertical temperature and salinity profiles were measured at more than 3000 stations and, although not taken strictly simultaneously, they provided a representative picture of the actual situation, in the summer of 1958, of the boundary zone between the warm, highly saline branches and extensions of the Gulf Stream in the south and the cold, low-salinity waters of polar origin in the north. This mixing of water masses represents a very important factor which influences the weather and consequently the fishing industry in Europe. The complicated structure of the Gulf Stream phenomenon with its meanders and eddies of warm and cold water spinning off, was revealed through these data.

Table 3 — Third Phase of Ocean Exploration
Research Ships Participating in the Atlantic Polar Front Survey 1958

Country	Ship	Country	Ship
Canada	Investigator Sackville	Iceland	Aegir
Denmark	Dana	Norway	Helland Hansen Johan Hjort G.O. Sars
France	Aventure Calypso Eveillé Emporté Le Verrier	UK	Discovery II Explorer Sarsia Vidal
Germany	Anton Dohrn Gauss	USSR	Ekvator M. Lomonosov Poljarnik Sebastopol

A similar co-operative effort was the so-called *Overflow Expedition* in 1960, aimed at studying the outflow of Arctic bottom water over the submarine sill between Iceland and the Faeroe Islands, which is of considerable significance for the renewal of the Atlantic bottom water. This programme, carried out by 9 research vessels from 5 European countries, was also organized by ICES.

Until recently, participation in such international co-operative work has been limited to countries sufficiently developed in the field of oceanography to be able to provide the necessary ships, scientific equipment and personnel.

Gradually though, interest in marine affairs, particularly in the nature and resources of the ocean, has been increasing in the developing world. Such interest, which gained strong impetus from the International Geophysical Year 1957/58, called for a new, more comprehensive and world-wide approach to oceanography. A new system of international co-operation had then to be developed : a machinery which would permit the participation of all nations on the globe that are interested in oceanography, and would promote world-wide co-operation and co-ordination, to the benefit of all mankind. Unesco entered the stage to play a role in marine research at this time.

5 The founding of IOC

Unesco's interest in oceanography was first documented in 1950 when the 8th session of its General Conference, held in Montevideo, Uruguay, authorized the Director-General to promote the co-ordination of research on scientific problems relating to a number of fields including oceanography and marine biology. This impetus eventually led to the formation of an *International Advisory Committee on Marine Science (IACOMS)* in 1955, composed of nine marine scientists appointed by the Director-General and chosen from among the members of an international panel of honorary consultants set up by the Unesco Secretariat, after consultation with the FAO Secretariat. The functions of IACOMS were to advise the Director-General of Unesco on the promotion of international collaboration in marine science in the preparation and execution of marine research projects, taking into account related programmes of the United Nations and other Specialized Agencies. IACOMS held several meetings in the following years and built up close collaboration with the *Special* (later renamed *Scientific) Committee on Oceanic Research (SCOR)*, a non-governmental body established by the *International Council of Scientific Unions (ICSU)*. SCOR, after its first meeting in 1957, was involved in stimulating, organizing and co-ordinating a co-operative effort to investigate the Indian Ocean. Owing to the seasonal change of the monsoons, this ocean holds a special position among the oceans. Its reaction on atmospheric forces, although of great economic importance to the countries in the region, was only poorly understood and, therefore, constituted a substantial challenge to marine scientists. Moreover, as this ocean is surrounded mostly by developing countries, a major international research effort in that area was envisioned as an encouraging and beneficial influence for the countries of the region. Unesco agreed to co-sponsor this programme, the *International Indian Ocean Expedition (IIOE)*, but it became more and more clear that, for such a vast international co-operative undertaking which required the provision of research ships and research facilities ashore, the commitment of governments would be necessary. Realizing the need for dynamic and co-ordinated intergovernmental action, Unesco's General Conference, at its 10th session in November 1958, decided to convene an intergovernmental conference on oceanographic research, which, after a preparatory meeting in Paris in March, was held in Copenhagen in July 1960. International organizations that were involved in the preparation of this conference were Unesco, the United Nations, FAO, WMO and IAEA. The principal recommendation of the Copenhagen conference was that an *Intergovernmental Oceanographic Commission (IOC)* be established within the framework of Unesco, with the task of promoting concerted action of Member States in the field of oceanographic research. This recommendation was adopted by the 11th session of Unesco's General Conference in November/December 1960 which also approved the Statutes of the IOC, and an *Office of Oceanography* was set up within the Unesco Secretariat, which was to operate as IOC's Secretariat. The necessary funds were accordingly allocated.

It is interesting to recall that this speedy and determined action was achieved in a time period when many oceanographers and politicians believed that mankind was at the start of an «oceanic age»,characterized by a highly significant and comprehensive development of ocean science and oceanic resources.

With the necessary prerequisites thus arranged, IOC held its first session in October 1961, and the organizational structure, the provisional rules of procedure, and the general directives for its further programme were decided upon. A Bureau composed of the Chairman and the two Vice-Chairmen was elected (see Table 4) and, in addition, a Consultative Council was established to work with the Bureau and the IOC Secretariat in the development of the programme of the Commission. This Consultative Council was put on a more formal basis by the third session which adopted appropriate changes of the IOC Statutes. These were approved by the 13th Session of its General Conference of Unesco in 1964. Several working groups were formed to deal with special subjects such as a network of fixed oceanographic stations, radio communications, exchange of oceanographic data. The Secretary was instructed to assume the co-ordinating functions of the International Indian Ocean Expedition, in consultation with SCOR. Furthermore, the Commission, in an attempt to make suitable co-operative arrangements with other international organizations, invited the United Nations and its Specialized Agencies as well as other intergovernmental and non-governmental organizations concerned with various disciplines of oceanography to co-operate with IOC to the fullest possible extent. In particular, IOC expressed the hope that the interested agencies of the UN family would find it possible to designate members of their Secretariats to co-operate actively with the IOC Secretariat. In addition, preparations for the establishment of advisory channels to the Commission in all fields of marine science were made. The tradition of scientific lectures was also started at this time by a talk in honour of Fritjof Nansen, the great Norwegian oceanographer and explorer of the Arctic Ocean (see Section 15).

Table 4 — Chairmen, Vice Chairmen and Secretaries of IOC (1961 - 1977)

Time of Election	Chairman		Vice Chairmen	Secretary
1961	Dr. A. Bruun (Denmark)		Dr. W.M. Cameron (Canada) Vice Adm. V.A. Tchekourov (USSR)	Dr. W.S. Wooster (USA)
1962	Dr. W.M. Cameron (Canada)		Captain L.R.A. Capurro (Argent.) Vice Adm. V.A. Tchekourov (USSR)	"
1964	Dr. N.K. Panikkar (India)		Prof. H. Lacombe (France) Prof. K. Sugawara (Japan)	Dr. K.N. Fedorov (USSR) from Sept. 1963
1965	Prof. H. Lacombe (France)		Commodore W. Langeraar (Netherlands) Prof. F. Pautsch (Poland)	"
1967	Rear Adm. W. Langeraar (Netherlands)		Prof. C. Morelli (Italy) Dr. J. Carranza Frazer (Mexico)	"
1969	"		Prof. C. Morelli (Italy) Captain O.A. Amaral Affonso (Brazil)	" Dr. S. Holt (UK) from Jan. 1970
1971	"	I	Prof. Dr. H.U. Roll (Fed. Rep. of Germany)	"
		II	Capt. O.A. Amaral Affonso (Brazil)	
		III	Prof. Dr. Niegolewski (Poland)	
		IV	Prof. Dr. K. Sugawara (Japan)	
1973	Dr. G.F. Humphrey (Australia)	I	Prof. Dr. H.U. Roll (Fed. Rep. of Germany)	Mr. D.P.D. Scott (UK) from Oct. 1972
		II	Dr. G. Serpoianu (Romania)	
		III	Dr. A. Ayala-Castanares (Mexico)	
		IV	Prof. Dr. S.K. El Wakeel (Egypt)	
1975	"	I	Dr. A. Ayala-Castanares (Mexico)	"
		II	Dr. G. Serpoianu (Romania)	
		III	Prof. Dr. S.K. El Wakeel (Egypt)	
		IV	Mr. O.J. Østvedt (Norway)	
1977	Dr. A. Ayala-Castañares (Mexico)	I	Dr. N.J. Campbell (Canada)	"
		II	Prof. Dr. C. Druet (Poland)	
		III	Mr. O.J. Østvedt (Norway)	
		IV	Capitán de Navío G. Angel Mejía (Colombia)	

6 Growth and development of IOC

After the promising start in 1961, the work of IOC grew and expanded in many respects. The Commission usually met every two years to decide on the main issues. In the interim period, the Bureau and Consultative Council, re-constituted in 1970 as the Executive Council, was convened every eight months in order to control and guide the work of the IOC Secretariat in implementing the decisions of the Commission (see Table 5).

The Commission attracted much interest, both in developed and developing countries, and the number of its members increased from 40 in 1961 to 103 in 1979 (see Fig 1). A list of the present IOC Member States is given in Annex I.

The Commission became aware of the need for the Statutes to be adapted to this growing and changed membership and to the increasingly more important role IOC was asked to play in marine scientific research. The sixth session of the Commission in 1969 accepted revised Statutes. The main changes were a re-formulation of the functions of the Commission and the replacement of the Bureau and Consultative Council by an Executive Council composed of the Chairman and 4 Vice-Chairmen as well as of representatives of not more than one fourth of the number of Member States, elected with due regard to the principle of geographical distribution. While, before, decisions were taken only by the Bureau after consultation with the Consultative Council, now such decisions are passed by the full body of the Executive Council. The revised Statutes were approved by the 16th Session of the General Conference of Unesco in 1970. A revision of the provisional Rules of Procedure followed.

Figure 1 - Development of IOC Membership 1961-1979 (March)

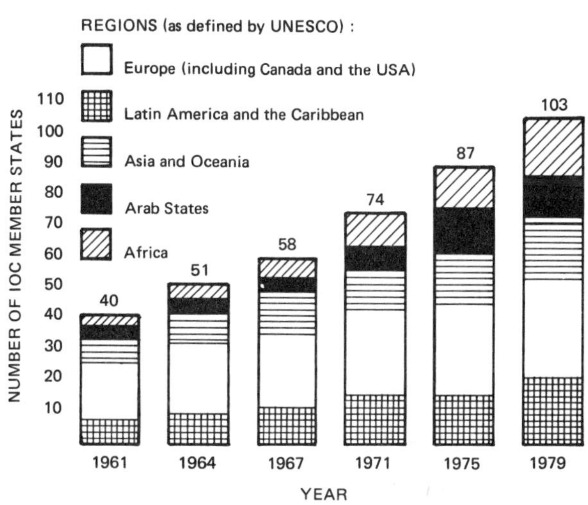

A short summary of the *functions* of the Commission as they are laid down in Article 2 of the Statutes of 1970 is given below.

The Commission shall :

- define those problems, the solution of which requires international co-operation in the field of oceanic research ; and
- develop, recommend and co-ordinate international programmes for
  - scientific investigations of the oceans,
  - related services,
  - strengthening education, training and assistance in marine science and its technology

which call for concerted action

- by IOC Member States
- with interested international organizations.

Upon first consideration, these functions seem to be primarily of a recommending or co-ordinating nature which require reaction upon initiatives from outside or inside the Commission and leave the implementation to its Member States or other international organizations. If examined more closely, however, it becomes evident that the expressions «define» and «develop» imply a substantial amount of potential initiative and operational freedom, provided that the Commission is aware of such responsibility and prepared to act accordingly.

How have these statutory requirements been transformed into practice ? The present activities of IOC, as they have been developed in the course of time, can be subdivided into three main areas :

- Ocean sciences, i.e. promotion and co-ordination of co-operative investigations in the fields of marine research ;
- Ocean services, i.e. organization of services to the scientific community and the public (data exchange, network of oceanographic stations, oceanographic products as e.g., analyses and forecasts of oceanic conditions, tsunami warning system) ;
- Education, training and mutual assistance (TEMA) which includes assurance to all interested IOC Member States that they can participate adequately in IOC's activities.

In order to utilize the expertise and capabilities of IOC's Member States, most of these functions are dealt with in the subsidiary bodies (e.g., Working Committees, International Co-ordination Groups, Working Groups, Groups of Experts) of the Commission, while the IOC Secretariat renders technical and operational assistance.

The activities of the Commission will be described in more detail in Sections 10 to 15.

Table 5 - Date and Place of sessions of the IOC Assembly, sessions of the IOC Bureau and Consultative Council, sessions of the IOC Executive Council from 1961 to 1978

Assembly			Bureau with Consultative Council		
Nº	Date	Place	Nº	Date	Place, Host
I	Oct. 1961	Paris, Unesco			
			I	April 1962	Paris, Unesco
II	Sept. 1962	" "			
			II	May 1963	Moscow, USSR
			III	Oct. 1963	Paris, Unesco
III	June 1964	" "			
			IV	Dec. 1964	Paris, Unesco
			V	June 1965	Rome, FAO
IV	Nov. 1965	" "			
			VI	May 1966	Paris, Unesco
			VII	Jan/Feb. 1967	Monaco, Principality of Monaco
V	Oct. 1967	" "			
			VIII	June 1968	London, UK
			IX	Febr. 1969	Woods Hole, USA
			X	Sept. 1969	Paris, Unesco
VI	Sept. 1969	" "			
			XI	Jan. 1970	Paris, Unesco
			XII	March 1971	Bordeaux, France
			XIII	Oct. 1971	Paris, Unesco
VII	Nov. 1971	" "			
			Executive Council		
			I	July 1972	Hamburg, FR Germany
			II	May 1973	Paris, Unesco
			III	Nov. 1973	Paris, Unesco
VIII	Nov. 1973	" "			
			IV	June 1974	Ottawa, Canada
			V	March 1975	Venice, Italy
			VI	Oct. 1975	Paris, Unesco
IX	Nov. 1975	" "			
			VII	June 1976	Bergen, Norway
			VIII	April 1977	Paris, Unesco
			IX	Oct. 1977	Paris, Unesco
X	Nov. 1977	" "			
			X	June 1978	Rome, FAO

CHAPTER II

Co-operation with other bodies

7 External influences on IOC' s functions and structure

Apart from the International Indian Ocean Expedition, which was already operational when the Commission was founded, most of IOC's widespread activities have developed from internal proposals made by one or more of its Member States.

The Commission at its first session in 1961 indicated the research areas which should be covered by co-operative programmes. At its second session in 1962, IOC requested SCOR to prepare a general scientific framework for the comprehensive study of the world's oceans to serve as a basis for developing world-wide international programmes for ocean investigation. It was printed as a draft in 1965 and later, in its final version, published as «Perspectives in Oceanography, 1968» in 1969 (*IOC Technical Series No 6*). Such listing of potential research areas, based on purely scientific motivation, however, does not constitute an international programme in itself, but is the scientific basis of such a programme. The decisive impetus to develop such a long-term programme originated from the *United Nations* ; this being an example of an action generated by an impulse from outside IOC.

In December 1966, the General Assembly of the United Nations, recognizing the need for greater knowledge of the oceans and the opportunities available for the utilization of their resources, requested the Secretary-General of the UN to make proposals with the aim of ensuring the most effective arrangements for an expanded programme of international co-operation. These proposals were to be made in co-operation with Unesco, in particular its IOC, and with FAO, «to assist in a better understanding of the marine environment through science and in the exploitation and development of marine resources» (resolution 2172 (XXI)).

When dealing with this UN resolution, IOC obtained valuable advice from a joint working group formed by its scientific advisory bodies (see Section 9), which met at Helio Cabala near Rome in the summer of 1967 and produced the report «International Ocean Affairs». This report, in addition to its scientific suggestions, contained a number of recommendations on organizational and financial matters aimed at improving and strengthening intergovernmental co-operation in the field of marine science and proposing, as long-term action, «that the member governments of the UN family and the various UN agencies give early and thorough consideration to the advisability and feasibility of establishing a *central intergovernmental oceanic organization* to deal with all aspects of ocean investigation and the uses of the sea».

At its fifth session in October 1967, the Commission, after having this report examined by the IOC Working Group on Intergovernmental Aspects of the Implementation of UN Resolution 2172, recognized the necessity of obtaining increased financial support but came to the conclusion that it was too soon to suggest any charges in the present organizational arrangements.

Meanwhile, in response to UN resolution 2172 (XXI) and assisted by a group of experts, among them the Chairman of IOC, the Secretary-General of the United Nations had pepared a comprehensive report (E/4487), dated 24 April 1968, which *inter alia* recommended

- that *an expanded programme* of international co-operation in marine research synthesizing national and international plans in this field be developed and formulated *by a suitable intergovernmental body* occupying a focal position and related to the international organizations concerned ; and
- that Member Governments, Unesco, FAO, WMO, and other UN organizations as may be concerned agree as a matter of urgency *to broaden the base of IOC* so as to enable it *to formulate and co-ordinate such an expanded programme*.

By resolution 2414 (XXIII), of 17 December 1968, the General Assembly of the United Nations endorsed this concept of a long-term and expanded programme of oceanographic research and requested that «a comprehensive outline of the scope of this programme be drawn up».

In the same session, the UN General Assembly adopted resolution 2467 (XXIII) which welcomed the concept of an *International Decade of Ocean Exploration* to be undertaken within the framework of the long-term and expanded programme and requested IOC to «intensify its activities in the scientific field, within its terms of reference and in co-operation with other interested agencies, in particular with regard to co-ordinating the scientific aspects of a long-term and expanded programme of world-wide exploration of the oceans and their resources of which the International Decade of Ocean Exploration will be an important element, including... international efforts to strengthen the research capabilities of all interested nations with particular regard to the needs of the developing countries».

Again IOC obtained advice from a joint working group of its advisory bodies which was convened at Ponza and Rome in the spring of 1969 and from which the report «Global Ocean Research» was conceived. Thereafter, a special IOC working group drafted the comprehensive outline of the scope of the *Long-term and Expanded Programme of Oceanic exploration and Research (LEPOR)* which was approved by IOC at its sixth session in September 1969. This document was subsequently forwarded to the Secretary-General of the UN for submission to the UN General Assembly which acknowledged it by resolution 2560 (XXIV) in December 1969. The General Assembly asked IOC to keep LEPOR up to date and to implement it in appropriate stages, in co-operation with other international organizations concerned.

At its sixth session, IOC further decided to establish a Group of Experts on Long-Term Scientific Policy and Planning (GELTSPAP) to keep this long-term and expanded programme under continual review and to advise the

Figure 2 - Intergovernmental Oceanographic Commission Working Structure

(as of 31 December 1978)

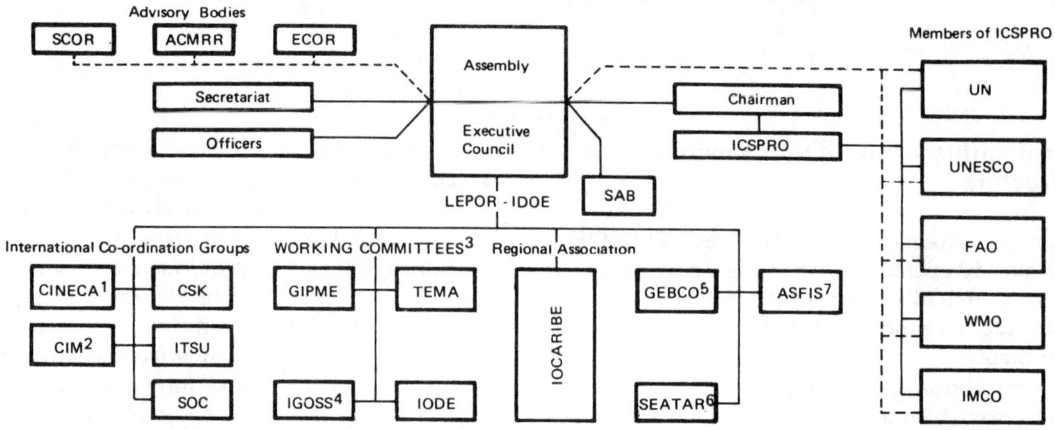

ACMRR Advisory Committee on Marine Resources Research (of FAO)
ASFIS Aquatic Sciences and Fisheries Information System (FAO/IOC)
CCOP Committee for Co-ordination of Joint Prospecting for Mineral Resources in Asian Offshore Waters (of ESCAP)
CECAF Fishery Committee for the Eastern Central Atlantic (of FAO)
CIM Co-operative Investigations in the Mediterranean (IOC/FAO(GFCM)/ICSEM))
CINECA Co-operative Investigation of the Northern part of the Eastern Central Atlantic (ICES/IOC/FAO(CECAF))
CSK Co-operative Study of the Kuroshio and adjacent regions
ECOR Engineering Committee on Oceanic Resources
ESCAP Economic and Social Commission for Asia and the Pacific (of UN)
FAO Food and Agriculture Organization of the United Nations
GEBCO General Bathymetric Chart of the Oceans (IOC/IHO)
GIPME Global Investigation of Pollution in the Marine Environment
GFCM General Fisheries Council for the Mediterranean (of FAO)
ICES International Council for the Exploration of the Sea
ICSEM International Commission for the Scientific Exploration of the Mediterranean
ICSPRO Inter-secretariat Committee on Scientific Programmes Relating to Oceanography
ICSU International Council of Scientific Unions
IDOE International Decade of Ocean Exploration
IGOSS Integrated Global Ocean Station System
IHO International Hydrographic Organization
IMCO Inter-Governmental Maritime Consultative Organization (of UN)
IOCARIBE IOC Association for the Caribbean and adjacent regions
IODE International Oceanographic Data Exchange
ITSU Tsunami Warning System in the Pacific
LEPOR Long-term and Expanded Programme of Oceanic exploration and Research
SAB Scientific Advisory Board
SCOR Scientific Committee on Oceanic Research (of ICSU)
SEATAR Studies on East Asia Tectonics and Resources (CCOP/IOC)
SOC Southern Oceans
TEMA Training, Education and Mutual Assistance in the marine sciences
UN United Nations
UNESCO United Nations Educational, Scientific and Cultural Organization
WMO World Meteorological Organization (of UN)

(1) Joint with ICES and FAO(CECAF)
(2) Joint with FAO(GFCM) and ICSEM
(3) The Working Committees and their subsidiary groups may have direct links with the relevant Advisory Bodies and/or their subsidiary bodies.
(4) The IGOSS is a joint programme with WMO
(5) Joint Guiding Committee
(6) Joint Working Group
(7) Joint Panel of Experts

Commission on its implementation. The International Decade of Ocean Exploration (IDOE), covering the period from 1971 to 1980, was considered an important element of LEPOR and its acceleration phase. Since 1969, LEPOR has formed the principal framework of the scientific activities of IOC. It was published as *IOC Technical Series* No 7 in 1970.

Another external influence on the development of the Commission originated from the *UN Conference on the Human Environment* held in Stockholm in 1972. This conference requested IOC to create a programme for the global investigation of pollution in the marine environment, though such a programme had already been initiated by IOC and was considered one of the major projects in IDOE, the acceleration phase of LEPOR. More important, the Stockholm conference, after stressing the need for supporting the marine environmental programme just mentioned, recommended that governments :
«expand their support to components of the UN system concerned with research and monitoring in the marine environment and adopt measures required *to improve the constitutional, financial and operational basis under which the IOC is at present operating* so as to make an effective joint mechanism for the governments and the UN organizations concerned... and in order that it may be able to take addditional responsibilities for the promotion and co-ordination of scientific programmes and services».

Deliberations on the structure of IOC, on improving its services, strengthening the relationships to other organizations and increasing the efficiency of its Secretariat had already been going on in the Commission. These were intensified and led to a *new organizational structure* of the Commission which was adopted by the eighth session of the IOC Assembly in November 1973 and is still in force (see Fig. 2). Prior to this restructuring, the Director-General of Unesco, following the request of the seventh session of the Commission in 1971, had reorganized, as from 16 October 1972, the IOC Secretariat and the Unesco Office of Oceanography (later renamed the Division of Marine Sciences) as separate entities, with the Secretary IOC reporting directly to the Assistant Director-General for Science.

8 Co-operation with other UN organizations and intergovernmental bodies

From its very beginning IOC has had to be aware of the fact that there are other international organizations of the UN family interested and engaged in certain areas of marine affairs. FAO deals with fisheries, WMO is concerned with marine meteorology, Unesco operates a programme of training, promotion of methodology and technical assistance in marine science and IMCO takes care of international aspects of shipping and safety at sea. The UN also has created an environmental programme which involves the marine environment. With a view to ensuring maximum progress and to avoiding duplication of effort, the Commission has paid much attention to establishing close relationships with those organizations and invited them to co-operate with IOC to the fullest possible extent.

In the course of time, the need for a formal basis of such a relationship was felt, and the Executive Board of Unesco at its 82nd Session, following relevant recommendations of the 9th Session of the IOC Bureau with the Consultative Council in February 1969, authorized the Director-General of Unesco to negotiate with other interested agencies of the UN system in order to establish a suitable mechanism for such co-operation. As the result of such negotiations the *Inter-secretariat Committee on Scientific Programmes Relating the Oceanography (ICSPRO)* was created in 1969. The objective of ICSPRO is to contribute to the development of effective forms of co-operation among organizations of the UN family concerned with oceanic programmes and, thus, to avoid duplication and overlapping in the planning and implementation of an expanded programme of international marine science, as desired by the UN General Assembly.

The Committee so far consists of the Executive Heads of the UN, FAO, Unesco, WMO and IMCO but is open to any other United Nations agencies which agree «to participate in the work of the Committee, and who have indicated that they are prepared, ..., to contribute to the Secretariat of the IOC, to sustain the work of the Commission through relevant parts of the programmes of their respective organizations, and to use the Commission as appropriate for advice and review in the area of marine science». The Chairman of IOC ex-officio «has the right to participate fully in sessions of ICSPRO».

Through this committee, the base of IOC has been broadened so that the Commission can fulfil its functions as an effective *joint specialized mechanism* for intergovernmental marine research. The members of ICSPRO provide support to IOC's activities in the form of co-operation in technical work, the provision of staff to the multi-agency Secretariat of IOC, conference services, publication facilities, and other means as needed. The ICSPRO members, «looking upon IOC as constituting a specialized part of their own machinery», refer to the Commission for advice and review and use the Commission in discharging certain of their responsibilities in the field of marine science. Likewise, the Commission may request the ICSPRO members to undertake the planning and implementation of those parts of the IOC's programmes which are of interest to their Member States.

This co-operation based on the ICSPRO agreement is functioning well to the benefit of the organizations concerned and to marine science in general. In 1974, approximately one quarter of the staff salaries and operational funds spent by the IOC Secretariat were provided for by the ICSPRO agencies and the UN Environment Programme (UNEP). This clearly documents that IOC, although established within Unesco, has a certain autonomy in carrying out its duties.

In this connection the respective functions of the IOC and of the Division of Marine Sciences of Unesco are of particular interest and importance. The main responsibilities of the Division are in the field of training, education and technical assistance. This is accomplished mainly by building up an appropriate national and regional infrastructure in marine science in Member States, particularly developing countries. IOC's task in this field is primarily of a co-ordinating nature, aimed at assisting Member States to participate in the scientific programmes of the Commission. For such purpose, IOC operates its own fund-in-trust and, since 1977, its Voluntary Assistance Programme (IOC-VAP). Since the two units, the IOC Secretariat and the Division of Marine Sciences of Unesco, are colocated in Unesco House in Paris, the contact necessary to avoid duplication and overlapping is easily achieved.

Apart from the global intergovernmental organizations with which co-operation has been formalized by the ICSPRO agreement, there are others that are of a regional character. The International Council for the Exploration of the Sea (ICES, already mentioned in Section 4) is engaged in promoting and co-ordinating marine research, chiefly directed to the benefit of fisheries, in the North Atlantic Ocean and its marginal seas, while the *International Commission for Scientific Exploration of the Mediterranean Sea (ICSEM)* operates more or less in the same way in the Mediterranean Sea. The *International Commission for the North-west Atlantic Fisheries (ICNAF)* deals with fishing aspects in the western part of the North Atlantic Ocean. That the IOC maintains good relationships with these regional bodies and others of similar kind not mentioned here, is demonstrated by the attendance and participation of representatives of these regional organizations at IOC meetings and by the establishment of joint working groups if needed.

9 Advisory bodies to the Commission

International co-operation among governments alone does not necessarily imply that positive results will be produced from the scientific point of view. Therefore, it is absolutely indispensable that sound scientific advice is provided at the very beginning and at every stage of co-operative investigations executed by governments and co-ordinated by intergovernmental agencies.

Basically, oceanographic research, like many other kinds of research, is best done by individuals or small groups of scientists working independently and driven only by their curiosity to find out what is unknown and to understand the secrets of nature.

On an international level, these scientists co-operate in non-governmental organizations where each country is represented by its most eminent scientists who act in their personal capacity. It is to such non-governmental international bodies that an intergovernmental organization like IOC must look if it wishes to obtain the best possible scientific advice available. IOC has done so from its start in 1961 and has enjoyed the services of four advisory bodies, namely the

- Scientific Committee on Oceanic Research (SCOR),
- Advisory Committee on Marine Resources Research (ACMRR),
- Advisory Committee on Oceanic Meteorological Research (ACOMR),
- Engineering Committee on Oceanic Resources (ECOR).

SCOR, established in 1957 by the International Council of Scientific Unions (ICSU), is a non-governmental organization supported by Unesco. This group initiated the International Indian Ocean Expedition (1959-1965), the co-ordination of which was later taken over by IOC when it was realized that governmental support and intergovernmental co-ordination were needed for such a comprehensive enterprise. SCOR, assisted by national institutions of the host countries, also organized the Joint Oceanographic Assemblies held in New York 1959, in Moscow 1966, in Tokyo 1970, and in Edinburgh 1976, which provided a general current survey of the state and results of oceanographic research. As a very active unit composed of the leading marine scientists of the world, SCOR has been able to respond to all requests regarding scientific advice from IOC by creating relatively small working groups of experts on the problem in question. In such a way, the various disciplines and research areas of oceanography may be covered by the scientific advice given or by the oceanographic programmes developed by SCOR. Thus, SCOR has, from the founding of IOC, taken an important part in IOC's work by rendering valuable scientific assistance. Detailed information thereon will be given in Section 10. Occasionally, useful ideas and proposals that have been brought forward within IOC were transferred to SCOR, because they could best be implemented by SCOR and its working groups. This was the case with the important work on intercomparison and standardization of measuring methods and techniques applied in oceanography, which was originally inspired by IOC but was delegated to SCOR because of the essentially scientific nature of this task. It is now carried out by special SCOR working groups. The most prominent contribution of SCOR to IOC was the draft of a «General Scientific Framework for World Ocean Study», requested by IOC to serve as guidance to the scientific programmes of the Commission and already mentioned in Section 7. The first draft appeared in 1965 and, after some re-editing, it was published as «Perspectives in Oceanography, 1968» in 1969 (*IOC Technical Series* No 6).

ACMRR is a more recently formed non-governmental body than SCOR, originally established to provide scientific advice to FAO. The Commission recognized the need for scientific advice on fishery oceanography and wished to ensure that the interest of fisheries as well as research resources provided for fisheries, are taken into account when oceanographic studies are planned and executed by IOC. Therefore, at its second session in 1962, IOC designated the ACMRR of FAO as the advisory body to the Commission on fisheries aspects of oceanography. Since some Member States of IOC (e.g. the USSR) were not members of FAO at that time, ACMRR, for the purpose of serving as a scientific advisory body to IOC, was augmented by two additional people selected from scientists active in the field of fishery oceanography in States not members of FAO.

When dealing with UN resolution 2172 «Resources of the Ocean» (see Section 7), IOC, in addition to the scientific advice from SCOR and ACMRR, felt the need to obtain such assistance also with regard to the meteorological aspects of IOC programmes. Such advice was first rendered by an Advisory Committee of the World Meteorological Organization which participated, together with SCOR and ACMRR, in the meeting at Helio Cabala and Rome (1967) which drafted the report «International Ocean Affairs» and later by the Advisory Group on Ocean Research of WMO which assisted with the preparation of the report «Global Ocean Research» at Ponza and Rome (1969). These reports represented a major effort of the three advisory bodies in response to the Commission's request for advice and were what led to the concept of IOC's long-term and expanded programme LEPOR. Later this meteorological advisory committee was renamed the *Advisory Committee on Oceanic Meteorological Research (***ACOMR***)* and was considered an official scientific advisory body to IOC. In 1976, however, WMO disbanded its ACOMR and replaced it by a new Executive Committee Panel of Experts on Meteorological Aspects of Ocean Affairs (EC/MAOA), composed of governmental representatives, and asked this panel to act as advisory body to IOC. At the same time, SCOR offered its services in the meteorological field to IOC. The Commission decided to seek meteorological advice from both of the bodies proposed and from any other competent scientific organization.

In 1969, the IOC Bureau and Consultative Council envisaged that the establishment of appropriate advisory channels regarding scientific aspects of ocean engineering may become necessary. The IOC, at its sixth session in 1969, recognizing the fact that engineering is relevant to data management, to the design, installation and maintenance of ocean data platforms and to many other aspects of LEPOR, encouraged the formation of an *Engineering Committee on Oceanic Resources* (**ECOR**) which subsequently was established by the international engineering community as a non-governmental organization. At its seventh session, the IOC accepted ECOR as an advisory body. ECOR's purpose is to serve as an international focus for professional engineering interests in marine affairs.

CHAPTER III

The scientific programme

10 Research programmes of IOC before LEPOR

To develop, recommend, and co-ordinate international programmes for scientific investigations of the oceans, which call for concerted action by its members, is the task of IOC according to its Statutes. Let us now deal with these activities in more detail, starting with the first decade of IOC's history from 1960 to 1970, the period before the long-term and expanded programme (LEPOR) was conceived.

10.1 International Indian Ocean Expedition (IIOE) 1959-1965

The IIOE was already under way when IOC appeared on the scene. Initiated and organized by SCOR, it was conceived as an exploratory programme to allow individual scientists to carry out their own specialized programmes of interest. Co-ordination in time and place of the various contributions to IIOE was not very strict. It consisted more of an exchange of information on the relevant national programmes than of a development and implementation of a joint research plan adopted and adhered to by all participants. When the IOC was formed, it provided helpful and effective arrangements, as a co-ordinating agency, on quite a number of occasions, including the following measures :

- establishment of an International Co-ordination Group for the IIOE composed of national co-ordinators for IIOE and dealing with data exchange, preparation of atlases, processing, analysis and publication of results ;
- arrangement of special custom facilities and courtesies for ships and personnel of the expedition ;
- publication of an IIOE Information Paper series ;
- establishment of International centres, e.g. the *Indian Ocean Biological Centre* (charged with sorting of zooplankton samples) at Ernakulam, South India, supported by Unesco and India ; the *International Meteorological Centre* at Bombay, supported by UNDP and WMO ;
- designation of a Fisheries Subject Leader for evaluation and co-ordination of the fisheries aspects ;
- arrangement for international standardization and intercalibration tests ; and
- agreement upon reference stations at 15 locations throughout the Indian Ocean for inter-comparison of methods and for information on seasonal changes.

The scientific advisory role on IIOE remained with SCOR which took care of the different disciplines by appointing small groups of appropriate experts.

Some of the research cruises carried out during IIOE are shown in Fig. 3, together with a list of participating Member States.

One of the most important issues of IIOE was the educational aspect, in so far as a substantial number of developing countries of the area became interested in marine science, obtained technical assistance in this field and developed national organizations to deal with international oceanographic co-operation.

Further, the two World Data Centres for Oceanography in Washington D.C. and Moscow were charged with collecting the data obtained during this international co-operative programme and Unesco accepted responsibility for the publication of the Collected Reprints of the IIOE, which were issued in eight volumes, together with an index.

The observational results of this great co-operative investigation have been summarized and published in the form of five comprehensive atlases :

- *IIOE Meteorological Atlas*, Vol 1 Surface Climate of 1963 and 1964, edited by C.S. Ramage, F.R. Miller and Charmian Jeffries, Washington, D.C. (1972) ; Vol.2 Upper Air, edited by C.S. Ramage and C.V.R. Raman, Washington, D.C. (1972) ;
- *IIOE Oceanographic Atlas*, edited by C. Wyrtki, Washington, D.C. (1971) ;
- *IIOE Phytoplankton Production Atlas*, edited by J. Krey and B. Babenerd, Kiel (1976) ;
- *IIOE Geological-Geophysical Atlas*, edited by G.B. Udintsev, Moscow (1975).

A verbal representation of the scientific results of the IIOE was given at the seventh session of the Commission in 1971 (see Section 15).

It is not possible here to present a detailed account of the many scientific results obtained by this international co-operation ; only a few of the more important findings are indicated, as follows.

The surface current regime in the northern Indian Ocean is influenced by the seasonally changing monsoon winds which blow strongly from the south-west in summer and gently from the north-east in winter. Through the observations during the IIOE it was discovered that the ocean does not react on the summer monsoon from the south-west by establishing a simple current gyre covering the main part of the northern Indian Ocean, but that this monsoon gyre appears to contain many relatively strong cyclonic and anticyclonic eddies, with dimensions in the range from 100 to 1000 km, capable of changing dramatically within two months or less (see Fig. 4). Numerical models were a great help in the understanding of these current features which certainly affect chemical and biological processes in the ocean.

The participants in the discussion of the biological results indicated that not even 0.1% of the primary production of the Indian Ocean was harvested by man's fishing at that time and that up to a tenfold increase in the fishing yield might be reached with present conventional means, which could be further augmented by new technology. IIOE provided the oceanographic basis for

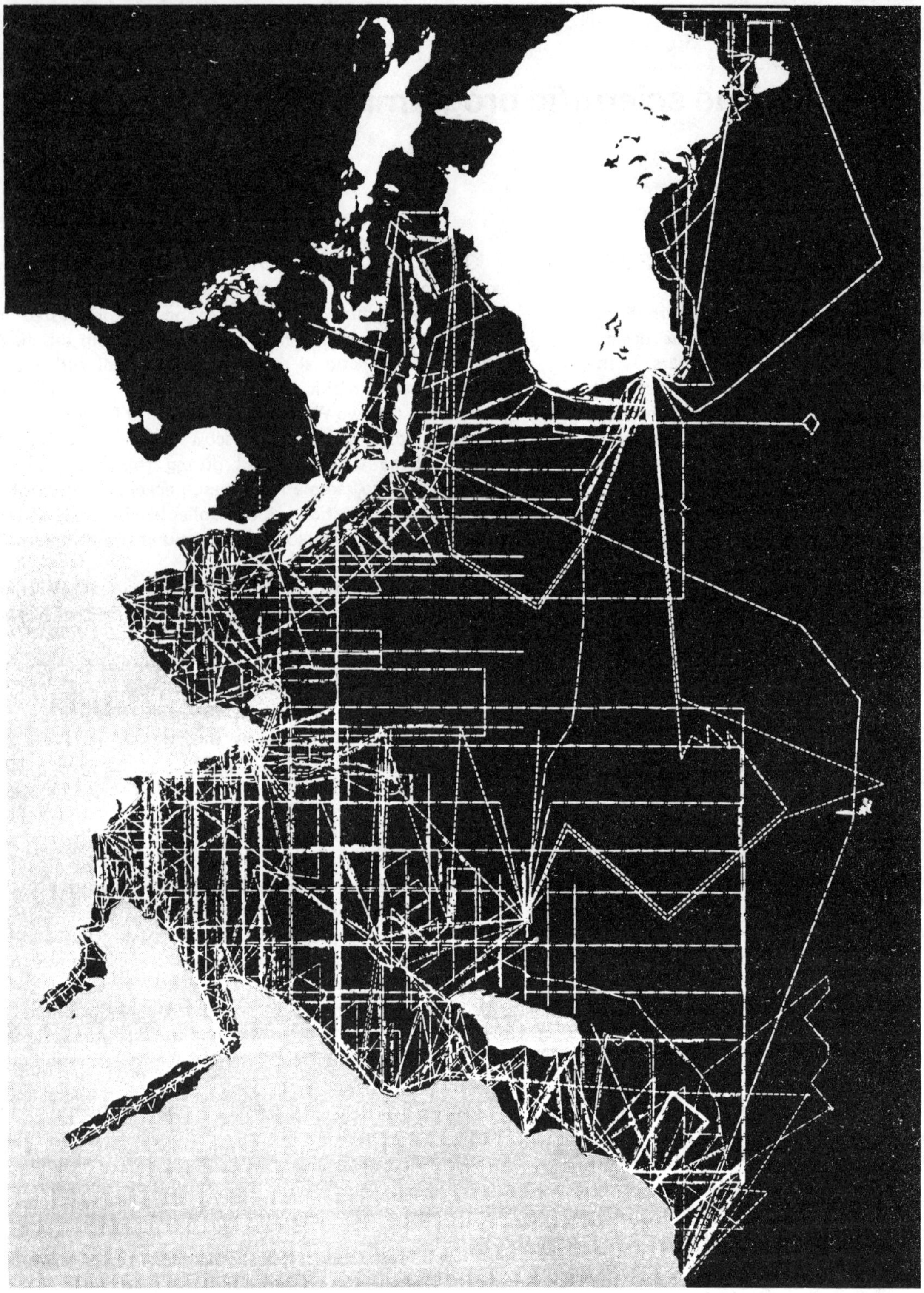

Figure 3 — Some of the cruises of the International Indian Ocean Expedition (IIOE), 1959-1965
(from Intergovernmental Oceanographic Commission (Five years of work) *IOC Tech. Ser.* No. 2, 1966).

Participating Member States

Ship-operating countries	*Number of ships*	*Other countries*
Australia	2	Burma
France	2	China
Germany (Fed. Rep.)	1	Ethiopia
India	2	Israel
Indonesia	1	Italy
Japan	5	Malagasy Republic
Pakistan	1	Fed. of Malaya
Portugal	1	Mauritius
Rep. of South Africa	4	Sri Lanka
Thailand	1	Sudan
United Kingdom	3	
USA	11	
USSR	5	
13 countries	**39 ships**	**10 countries**

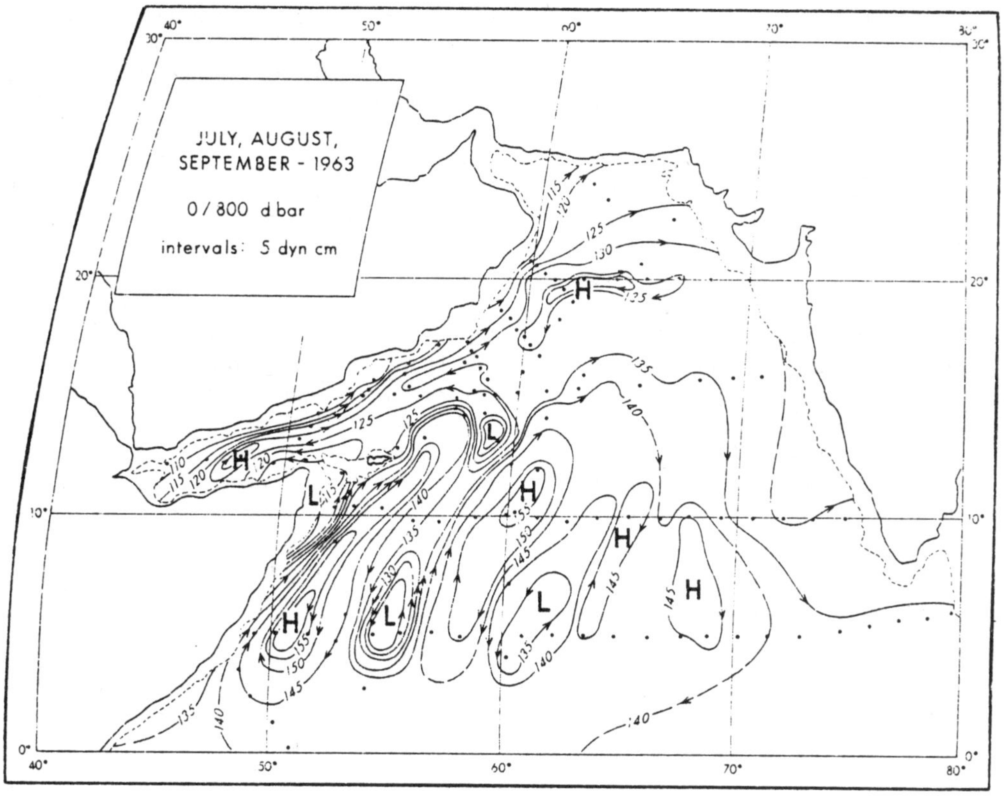

Figure 4 - The dynamic topography of the sea surface in the Arabian Sea during the summer monsoon of 1963 showing cyclonic and anti-cyclonic current eddies (From W. Düing, The monsoon regime of the currents in the Indian Ocean: International Indian Ocean Expedition Monographs Number 1 - East-West Center Press, Honolulu, Hawaii, 1970).

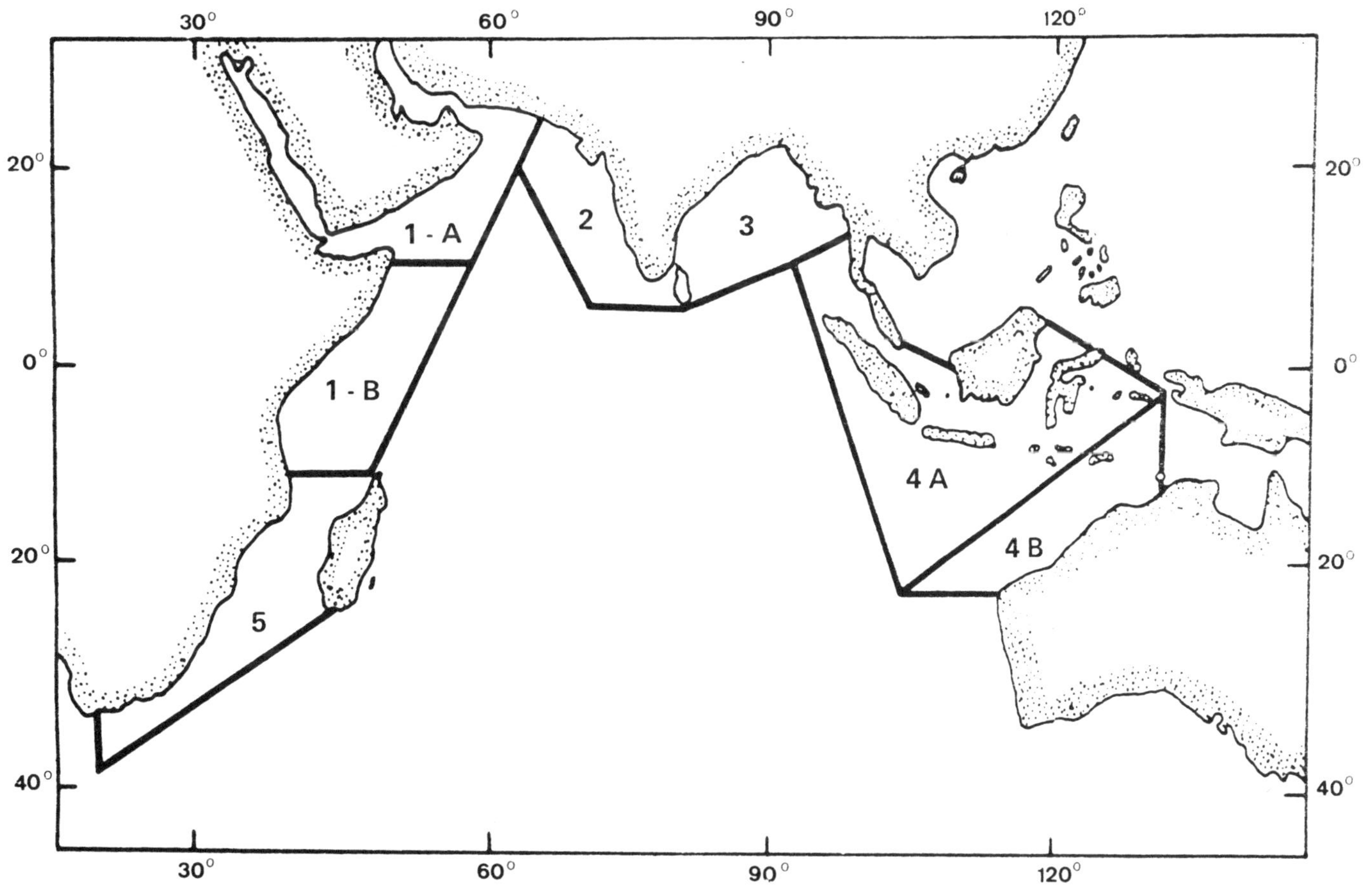

Figure 5 - Potential development areas for fisheries in the Indian Ocean (From G.F. Humphrey, The biology of the Indian Ocean. In: Bruun memorial lectures. *IOC Tech. Ser.*, No. 10, 1972, pp. 7-22).

planning a rational exploitation of living resources. From oceanographic considerations, the most promising areas for development appeared to be Somalia, South Arabia, Malabar, Madagascar, and Java (see Fig. 5).

In the geological-geophysical field, the atlas mentioned above gives ample information, including relief maps of the ocean floor ; charts showing the depth of the sedimentary layer and of bedrock outcrops ; the deep structure of the earth's crust and the upper mantle, as revealed by seismic investigation ; magnetic and gravitational anomalies and many other relevant observations which are of paramount significance for current knowledge and further development of the concepts of plate tectonics and sea-floor spreading, all of which adds to our understanding of the history of the oceans. Among the discoveries was that of a «hot spot» of anomalously hot water trapped in a deep basin in the Red Sea. The locally rich metaliferous sediments found with this hot brine spring may be worth exploiting.

10.2 International Co-operative Investigations of the Tropical Atlantic (ICITA), 1963 - 1964

ICITA was the first co-operative programme initiated, organized and implemented by IOC. At its second session in 1962, the Commission, following a proposal of the United States, adopted the International Co-operative Investigations of the Tropical Atlantic as an official IOC programme and established the International Co-ordination Group for ICITA chaired by an International Co-ordinator. The ICITA programme can be considered an extension of an earlier proposal, called the *West African Guinean Year*, and had been contained in a list of programmes strongly recommended for consideration by Member States at the first session of the Commission in 1961. The Guinean Year aimed at a full inventory of the marine living resources and the environmental conditions in the Gulf of Guinea.

ICITA comprised an oceanographic multiple-ship survey of the tropical Atlantic Ocean between latitudes 18° N and 18° S from the west coast of Africa to South America. It was subdivided into three field periods Equalant I, II, and III ; Fig. 6 provides information on the field work and the participating countries. In addition, scientists from the Federal Republic of Germany, the United Kingdom and Venezuela participated onboard several of the vessels enumerated in Fig. 6.

The observations made included direct current measurements, meteorological and oceanographic (physical and chemical) as well as marine biological data, the latter comprising biomass, rate of primary production and zooplankton tows. Bathymetric and geophysical data could be collected from certain vessels. Data reports were compiled by the US National Oceanographic Data Center in Washington, D.C. and printed by the hydrographic services of U.S.A. and Argentina.

Figure 6 - Some of the cruises carried out during the Internationl Co-operative Investigations of the Tropical Atlantic (ICITA), 1963 - 1964
(from Intergovernmental Oceanographic Commission (Five years of work). *IOC Tech. Ser., 1966.*

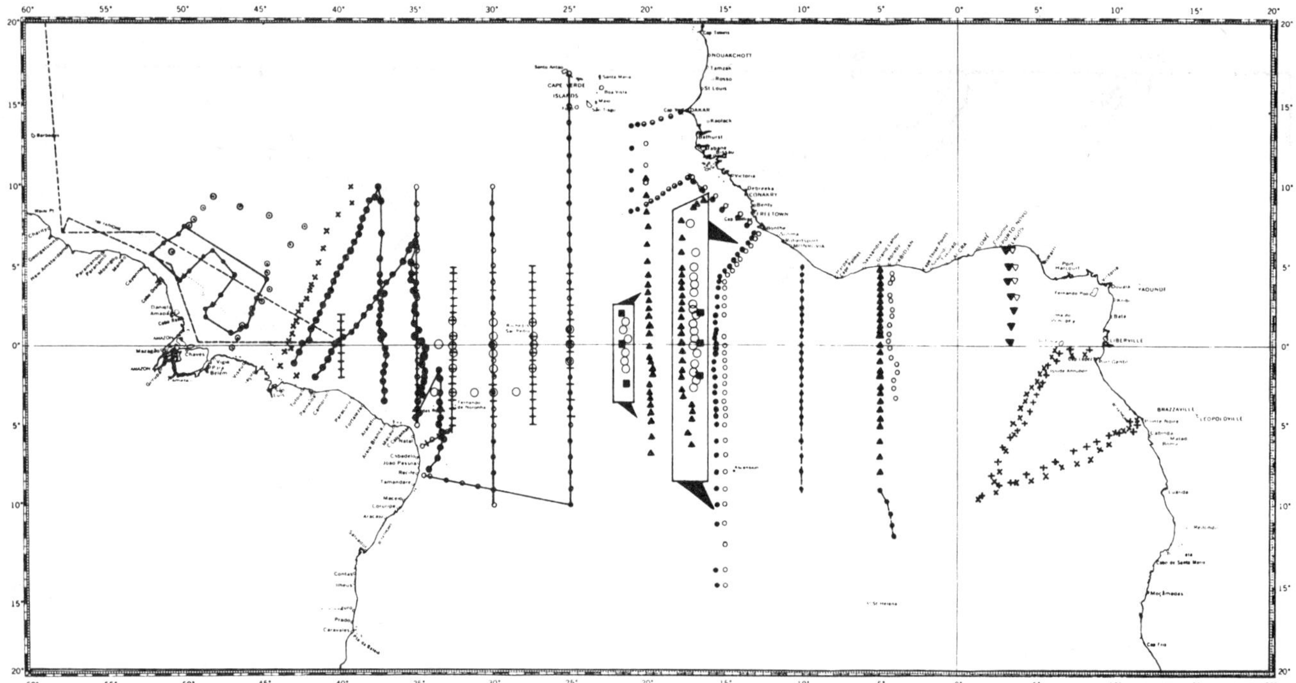

Participating Member States

Field periods	*Name of participating countries*	*Number*	*Number of research ships*
EQUALANT I February-April 1963	Argentina, Brazil, Rep. of Congo, Rep. of Ivory Coast, Nigeria, USA, USSR	7	14
EQUALANT II August-September 1963	Argentina, Brazil, Rep. of Congo, Rep. of Ivory Coast, Nigeria, Spain, USA, USSR	8	11
EQUALANT III February-March 1964	Rep. of Congo, Ghana, Rep. of Ivory Coast, Spain, USA, USSR	6	8

Following a relevant recommendation of the International Co-ordination Group of ICITA, the third session of the Commission in 1964 adopted the plan for the preparation of ICITA atlases, by an editorial committee under the chairmanship of the International Co-ordinator for ICITA. These atlases, composed of :

Vol. I Physical Oceanography — edited by A.G. Kolesnikov (USSR) ; and

Vol. II Chemical and Biological Oceanography — edited by A.G. Kolesnikov (USSR) and L.R.A. Capurro (Argentina), were published by Unesco in 1973 (Vol. I) and in 1976 (Vol. II). The results of ICITA and of the related Guinean Trawling Survey, sponsored by the Organization of African Unity, were also reviewed at a symposium in Abidjan, Ivory Coast in 1966.

ICITA was a remarkable step forward as far as co-ordination of such international investigations are concerned. The International Co-ordination Group met during the individual phases of the exercise and, thus, could finalize the plans for the following phase on the basis of the foregoing one. ICITA stands as a good example of a well-organized international co-operation.

A verbal report on the scientific results of ICITA was given at the eighth session of the IOC Assembly in 1973 (see Section 15).

Here again, only a few features of the physics, biology and geology that have been investigated in the tropical Atlantic Ocean are indicated.

In the physical field, The Atlantic Equatorial Undercurrent (see Fig. 7) was measured as early as 1911, although not «discovered» as such at this time. This undercurrent plays a very important role in equatorial circulation which is not yet fully understood.

Primary production is highest in coastal upwelling zones and in an oceanic area near the equator. Since plankton must be regarded as the basis of the food supply of fish, directly or indirectly, the spatial distribution of total plankton is of paramount importance in the search for areas with a high standing stock of fish. Biological observations during ICITA showed that the plankton distribution is characterized by a pronounced inhomogeneity or patchiness which, for a meaningful interpretation, would call for permanent registration of biological parameters at a network of stations over a longer period, supplemented by aerial photography.

Figure 7 - Scheme of the Equatorial Undercurrent in the Atlantic Ocean according to Kolesnikov et al., 1971, (from K. Voigt, The Atlantic Equatorial Undercurrent. In: Bruun memorial lectures, 1973. *IOC Tech. Ser.*, No. 11, 1975, pp. 12-19).
E = current setting east, W = current setting west, figures = current velocity in cm/sec.

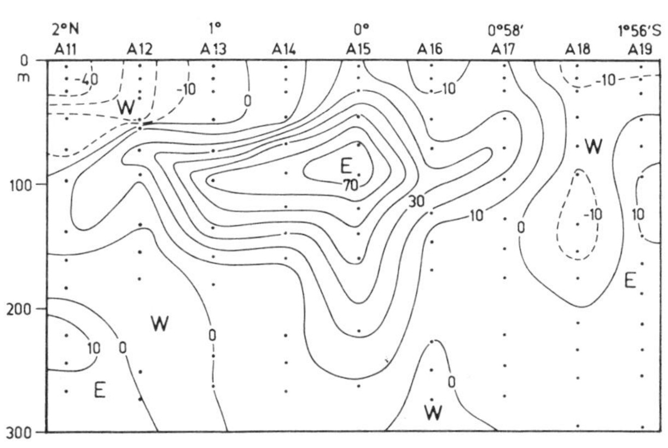

ICITA as well as the *Guinean Trawling Survey* (GTS) have greatly contributed to the increase of our knowledge on fishery resources in the Gulf of Guinea, although more accurate statistics covering catch, effort, size and composition of fish stock are still needed. Population dynamics models for the complex multi-species fisheries in the tropical seas are also lacking.

The exploration of the eastern Atlantic continental margin, carried out later than the field period of ICITA, revealed the presence of large basins showing a thickness of sediments of more than 4 km in a belt associated with the Congo and Niger rivers off Angola, Zaire, Cabinda, Congo, Gabon, Equatorial Guinea, Cameroon, Nigeria, Dahomey, Togo and Ghana (see Fig. 8). These sediments contain many structural features suitable for oil and gas traps and, owing to their location off major rivers, they are likely to have a high content of organic matter, which is the starting substance for the formation of oil and gas.

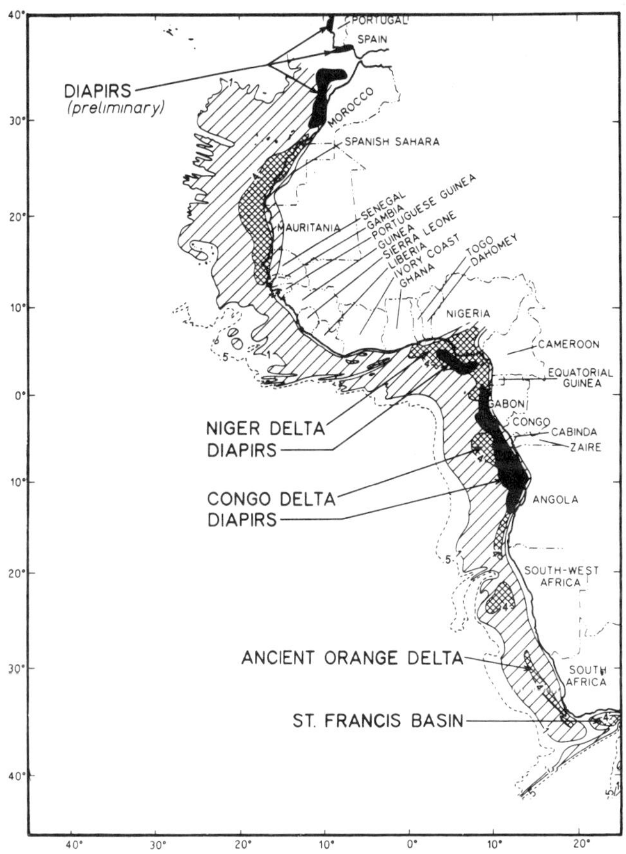

Figure 8 - Thickness of total sediment above continental and oceanic basement expressed in km. Areas of major promise for petroleum resources are indicated (from K.O. Emery, Review of the results from the Eastern Atlantic Continental Margin Programme of the International Decade of Ocean Exploration. In: Bruun memorial lectures, 1973. *IOC Tech. Ser.,* No. 11, 1975, pp. 52-62).

10.3 Co-operative Study of the Kuroshio and adjacent regions (CSK), 1965 - 1977

Similar to the Gulf Stream in the North Atlantic Ocean, the Kuroshio, flowing along the east coast of Asia, represents the most important current system in the North Pacific Ocean, exerting a far-reaching influence on climate, shipping, fisheries and the general economy in the whole of East Asia. Therefore, it is understandable that the significance of a thorough scientific investigation of the Kuroshio through international co-operation was recognized soon after IOC came into being. The idea of such a co-operative study originated from the Second Regional Meeting of Marine Science Experts in South-east Asia, convened by Unesco in Manila in 1962, where a relevant proposal was made by Japanese oceanographers. The IOC, at its second session in September 1962, approved appropriate action to be taken and recommended that interested Member States develop the programme for the co-operative seasonal investigation of the Kusoshio in the East China Sea, the southern Sea of Japan, and the eastern Philippines Sea (in 1965 the study was extended to the South China Sea). The investigation was to be comprised of studies of the physical, chemical, and biological oceanographic conditions and processes and also include fisheries oceanography and meteorological observations. Following this decision, a meeting of marine science experts was convened in Tokyo in October 1963, at which the present knowledge of various aspects of the Kuroshio was summarized and basic plans for the Co-operative Study of the Kuroshio and adjacent regions were formulated. After some re-consideration of the basic outline had taken place with regard to a more thorough development of fisheries aspects proposed by ACMRR, the third session of the IOC in 1964 adopted the Co-operative Study of the Kuroshio and adjacent regions (CSK) as an official IOC Programme and established an International Co-ordination Group composed of national co-ordinators nominated by participating Member States and chaired by an International Co-ordinator. This group was given the task of co-ordinating national efforts into a well-organizated exercise.

IOC further pointed to the necessity of including into the preliminary phase of CSK the following activities :

- analyses of existing data,
- preliminary field studies in key areas,
- field and laboratory tests of methods and equipment,
- training of personnel,
- strengthening of shore and ship facilities, and
- design of a field programme.

With a view to covering fisheries aspects, IOC later appointed an Assistant International Co-ordinator for Fisheries.

The first stage of CSK started in July 1965 ; 40 research vessels from 7 countries participated. Since then synoptic and multi-disciplinary surveys of the whole Kuroshio system have been carried out at least twice a year. The International Co-ordination Group has met eleven times to examine CSK activities and to guide its future work. Particular attention was paid to the standardization and intercalibration of measuring methods and techniques and reference stations were determined for the purpose of checking those methods.

The countries taking part in CSK are indicated in Fig. 9 ; some of the cruises are also shown. The data obtained were collected by the Japan National Oceanographic Data Center, functioning as the «Kuroshio Data Center». It took care of processing and publishing these data (16,727 oceanographic stations from 435 cruises up to 1975) in the form of «Data Reports of CSK» (328 volumes up to 1975) and *CSK Atlases* (7 volumes by 1977). The same centre also issued the «CSK Newsletter» in order to provide information about CSK activities. For preserving and sorting CSK biological, and particularly zooplankton, samples a Regional Biological Centre was established in Singapore in 1968.

The results of CSK were discussed at symposia held in Hawaii (1968), Tokyo (1970), and Bangkok (1973). The proceedings were later published under the titles «The Kuroshio», «The Kuroshio II» and «The Kuroshio III». An additional evaluation of CSK results took place at the ninth session of the IOC Assembly in 1975, where the Bruun Memorial Lectures were presented on the theme of the CSK (see Section 15).

The chief objective of CSK has been the study and understanding of the Kuroshio Current and its variations in time and space, with the main results being in the field of physical oceanography, both descriptive and model-oriented. Data accumulated by regular cruises of research vessels have provided a basis for studying seasonal and year-to-year variability in the oceanographic conditions of the CSK area and have, in addition, augmented previous knowledge of the biological productivity of those waters, thus rendering service to fisheries oceanography. In this respect, CSK has also greatly contributed to increasing the accuracy of the determination of the content of nutrients and other important substances in sea water by using CSK standard solutions developed by Japanese marine chemists.

New insight has been obtained into the dynamics of the Kuroshio which, in the same way as the Gulf Stream, is characterized by nearly stationary or moving meanders and cyclonic and anticyclonic eddies developing out of the meanders, or generated by typhoons, and sometimes having life-spans of about a year. A powerful tool for explaining and understanding the complicated dynamics of the Kuroshio was provided by numerical modeling of oceanic circulation processes. By applying such models, theoreticians tried to find out why the current takes a particular path, why it varies, and why the meanders may remain stable within certain boundaries for quite a long time. The peculiar bimodal path structure shown in Fig. 10 may be explained by influences originating from the earth's sphericity, its diurnal rotation, the shape of the ocean bottom, and the form of the vertical current profiles.

During the field phases, CSK obtained effective support from ICSPRO organizations. FAO and its ACMRR rendered advice and assistance with regard to carrying out fisheries studies as an important part of CSK, and facilitated the work of the CSK Assistant International Co-ordinator for Fisheries. Unesco provided shipboard fellowships as well as standard oceanographic equipment to developing countries so that they could participate effectively in the CSK programme.

In 1973, with the end of CSK approaching, considerations on the future programmes and also the future structure of CSK were brought forward by the International Co-ordination Group for CSK. The IOC, at the eighth session of its Assembly, requested that group to consider the type of mechanism needed for future co-operation in marine science in East Asia and to come up with appropriate proposals. Later, at its ninth session in 1975, the IOC Assembly decided that an *ad hoc* group of

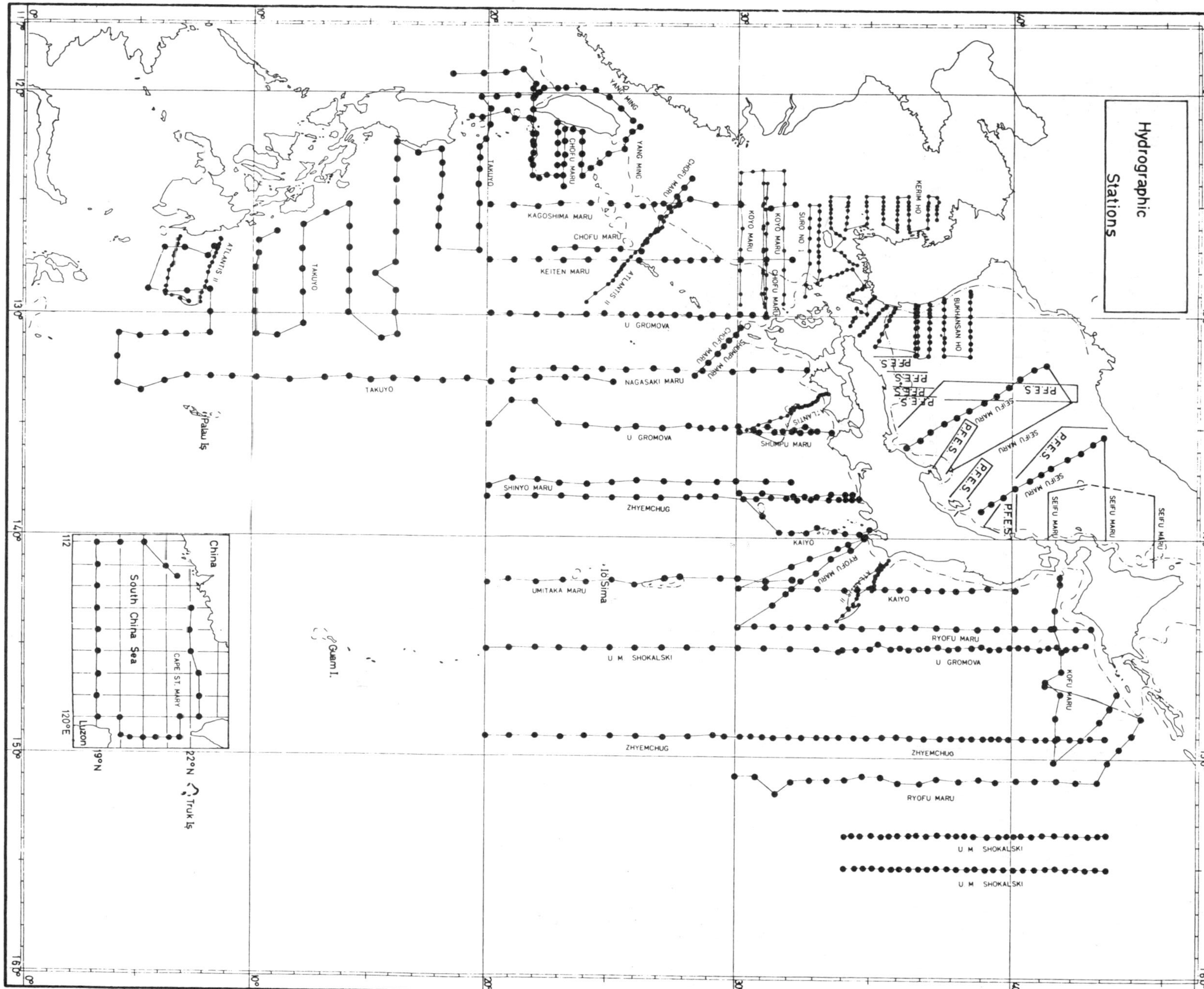

Figure 9 - Some of the cruises carried out during the Co-operative Study of the Kuroshio and adjacent regions (CSK), 1965-1977 (provided by Japan Oceanographic Data Center).

Participating Member States

Name of country	*Number of research ships*
First phase 1965-1966	
Japan	27
Rep. of Korea	3
Philippines	2
Rep. of China	1
UK (Hong Kong)	1
USA	3
USSR	3
	40
Later	
France	
Indonesia	
Japan	
Rep. of Korea	
Philippines	
Singapore	
Thailand	
UK (Hong Kong)	
USSR	
Soc. Rep. of Vietnam	

experts should be convened in order to define future programmes and co-operation in the western Pacific Ocean. Such a group was set up by the IOC Executive Council in 1976. Based on its report, the IOC Assembly, at its tenth session in 1977, finally decided to establish the *Working Group for the Western Pacific (WESTPAC)* which is open to all interested Member States. The activities of this working group cover a much larger area of the western Pacific Ocean than that studied by CSK. The scientific programmes envisaged are also on a much larger scale than CSK, and include, in addition to investigating the physico-chemical structure of the ocean and its dynamics, the problems of marine ecosystems and the productivity of the sea as well as geological and geophysical studies on the western margin of the Pacific Ocean. Marine pollution will be considered as well. Further, appropriate provision has been made for the aspects of training, education and mutual assistance. The establishment of a full-time WESTPAC Secretariat, located in the region, is planned. It is expected for WESTPAC to co-operate closely with other international bodies in the region, as e.g. with the Committee for Co-ordination of Joint Prospecting for Mineral Resources in Asian Offshore Areas (CCOP) and the Committee for Co-ordination of Joint Prospecting for Mineral Resources in South Pacific Offshore Areas (CCOP/SOPAC), both of the Economic and Social Commission for Asia and the Pacific (ESCAP). The CSK as a co-operative investigation of IOC will be terminated by the fourth Kuroshio Symposium, scheduled to take place in February 1979.

The establishment of WESTPAC must be considered an important step directed towards the regionalization of IOC, which implies setting up bodies charged with programmes of intergovernmental co-operative marine research in certain regions and serviced by a regional secretariat. The desire of the Member States in the region to continue co-operation on a permanent basis, clearly shows that interest in marine research among those countries has been substantially stimulated by co-operative investigations organized by IOC.

10.4 Co-operative Investigations of the Caribbean and Adjacent Regions (CICAR), 1967 - 1976

The idea of starting a co-operative investigation of the Caribbean Sea was put forward by the Netherlands in 1966. It was met with interest and encouragement by Member States, both in and outside the region, and by the scientific advisory bodies. Subsequently, the IOC, at its fifth session in 1967, adopted this proposal as an official IOC programme and established the International Co-ordination Group for the *Co-operative Investigations of the Caribbean and Adjacent Regions (CICAR)* composed of the National Co-ordinators of the various interested Member States plus representatives of FAO, WMO and other interested international bodies. 14 countries participated in CICAR (see Table 6).

The further development proceeded more or less along the lines prescribed by former co-operative enterprises :

- the international Co-ordination Group for CICAR met at regular intervals in the countries of the region in order to evaluate and guide current and planned research activities ;
- the National Oceanographic Data Center of the U.S.A. served as the Regional Data Centre for CICAR ;
- a sorting centre for receiving and processing biological samples and maintaining an International Reference Collection was established by Mexico with assistance of Unesco ;
- an Operations Centre for co-ordination and liaison purposes during the field operations of CICAR was set up in Curaçao and staffed with a Dutch marine scientist ;
- Unesco and FAO provided assistance, financially as well as in kind (standard oceanographic equipment, shipboard fellowships), to developing countries actively participating in CICAR ; and
- two CICAR International Symposia were held to discuss and evaluate the scientific results obtained.

Figure 10 - Paths of the Kuroshio Current for the period 1956 - 1964: A - Normal paths, B - Meander (abnormal) paths

(from A. R. Robinson, Dynamics of the Kuroshio current: experimental and theoretical studies from South to Kyushu to the Izu-Ogasawara ridge. In: Bruun memorial lectures,1975. *IOC Tech. Ser.,* No. 15, 1976, pp. 23-- 33).

A

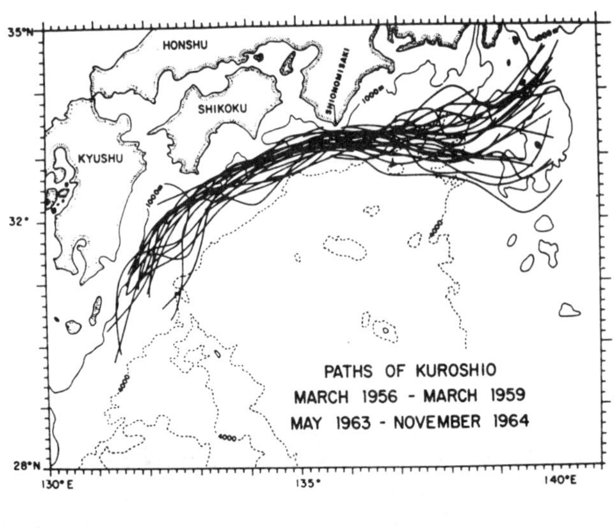

B

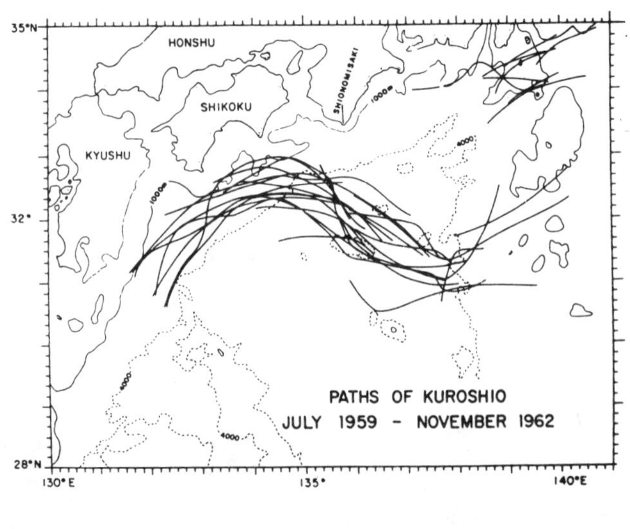

Table 6
Member States participating in the Cooperative Investigations of the Caribbean and Adjacent Regions (CICAR), 1967 - 1976

Brazil	Netherlands
Colombia	Panama
Cuba	Trinidad and Tobago
France	Union of Soviet Socialist Republics
Guatemala	United Kingdom
Jamaica	United States of America
Mexico	Venezuela

Unlike CSK, where the Kuroshio Current was the dominating subject to be studied, CICAR represents an area-oriented research programme with objectives more difficult to define than those of a problem-oriented investigation. The CICAR Survey Months Programme, which was the consequence of this situation, certainly contributed to CICAR's data output and data exchange, two of the very positive achievements of this co-operative investigation.

The most important effect of CICAR, however, is that many Caribbean countries became interested in marine affairs, and tried to create national marine science structures which constitute a necessary prerequisite for any further development in oceanography. Thus, CICAR's achievements appear to be more in the area of marine policy than in that of marine science. This is particularly exemplified by the wish of the Member States in the Caribbean region to continue co-operative marine research in an adequate form. Consequently, the IOC Assembly, at its ninth session in 1975, «decided to disband the International Co-ordination Group for CICAR and to establish, on an experimental basis for a period of six years, an *IOC Assocation for the Caribbean and adjacent regions (IOCARIBE)* for the purpose of continuing and developing regional co-operation in marine sciences built up over a period of seven years under the CICAR». Under the overall supervision of IOC, IOCARIBE is responsible for overseeing all activities in IOC in its region. It is developing a regional programme of marine science directed towards the needs of the countries in the area. The present membership comprises 16 countries from the Caribbean region and three from outside.

Again an experimental though decisive step with regard to regionalization was taken. CICAR was terminated in July 1976 by a symposium on the topic of «Progress in Marine Research in the Caribbean and Adjacents Regions» followed by the final session of the International Co-ordination Group for CICAR and the first session of IOCARIBE.

10.5 Other programmes

Apart from the scientific investigations described above, which required a substantial effort on the part of IOC, the Commission has kept under constant review other possibilities of rendering assistance to its Member States. Areas where such help was particularly welcome were the Mediterranean Sea and the waters around Antarctica.

An enquiry undertaken by the IOC Secretariat among IOC Members resulted in a consensus that a co-operative investigation in the Mediterranean was desirable. At its fifth session in 1967, the IOC, taking into consideration the scientific outline for such a programme presented by the USSR, resolved to adopt the *Co-operative Investigations in the Mediterranean (CIM)* as an official IOC programme starting from 1969 to be executed jointly with the International Commission for the Scientific Exploration of the Mediterranean Sea (ICSEM) and the General Fisheries Council for the Mediterranean (GFCM) of FAO. The necessary co-ordinating machinery was established in the form of an IOC/ICSEM/FAO(GFCM) International Group for Scientific Co-ordination and of a Group for Technical Co-ordination (GTC), composed of representatives of the IOC, ICSEM and GFCM Secretariats. Further, an Operational Unit was created and located at Monaco to assist the International Co-ordinator for CIM in the fulfilment of his task, e.g., by publishing a CIM Newsletter. The World Data Centre B for Oceanography in Moscow assumed the task of the Regional Data Centre for CIM. A Scientific Operational Programme for CIM was prepared and 24 IOC Member States from inside and outside the Mediterranean region declared their willingness to participate in this undertaking (see Table 7).

Table 7
Member States Participating in the Co-operative Investigations in the Mediterranean (CIM), 1969 -

Algeria	Monaco
Austria	Morocco
Egypt	Poland
Belgium	Romania
Bulgaria	Spain
France	Switzerland
Germany (Fed. Rep. of)	Syria
Greece	Tunisia
Israel	Turkey
Italy	Union of Soviet Socialist Republics
Lebanon	United Kingdom
Malta	Yugoslavia

There are various reasons why CIM has not achieved as much as had been hoped. Nevertheless, some success has been attained in certain special areas such as marine pollution, which was the subject of a scientific workshop held in Monaco in 1974, leading to a number of useful pilot projects on marine pollution in the Mediterranean. These were incorporated in the Mediterranean Action Plan of the United Nations Environment Programme (UNEP). Also, the amount of oceanographic station data has increased by 20%. Another noteworthy achievement was the preparation of a new International Bathymetric Chart of the Mediterranean (IBCM) on a scale of 1 : 1 million which reflects the scientific results obtained so far in this respect.

As regards oceanographic research in Antarctic waters, the Commission, at its fifth session in 1967, recognized the need for comprehensive studies in those regions expressed by the USSR and decided to establish an *IOC International Co-ordination Group for the Southern Oceans* (SOC), to provide effective co-ordination of oceanographic investigations carried out by countries adjacent to that aera and from the Northern Hemisphere. This action was aimed at the gradual evolution of a comprehensive study of the Southern Oceans. Although not much happened following this IOC action, the IOC International Co-ordination Group later gained importance with the new programme for the Biological Investigation of Marine Antarctic Systems and Stocks (BIOMASS), proposed by the Scientific Committee on Antarctic Research (SCAR) and SCOR. IOC's contribution to that programme is to facilitate data exchange as well as to collect and disseminate information on scientific programmes and ship schedules.

IOC's attention was further drawn to a study of the living resources and environmental conditions in the Northern part of the Eastern Central Atlantic, extending from the Strait of Gibraltar to Cap Verde. The need was felt for a synoptic study of the oceanographic conditions on the north-west African shelf with particular emphasis on upwelling conditions, bringing water rich in nutrients to the sea surface, on plankton distribution and fisheries. It was suggested that such a co-operative study might be organized jointly by IOC, ICES and the Fishery Committee of the Eastern Central Atantic (CECAF) of FAO. The IOC, at its sixth session in 1969, adopted the *Co-operative Investigations of the Northern part of the Eastern Central Atlantic (CINECA)*, in principle, as a co-operative programme of the Commission, but left the function of leading agency to ICES. IOC's role in CINECA as a co-sponsoring agency was limited to urging its Member States to make research ships available for the multi-ship surveys, to collect sea surface temperatures on board vessels operating in the CINECA area, and to respond fully to other requirements brought forward by the International Co-ordinating Group for CINECA.

11. Long-term and Expanded Programme of Oceanic exploration and Research (LEPOR)

11.1 LEPOR as a comprehensive framework for ocean research

The development leading to LEPOR was described in Section 7. Let us now consider the contents of this comprehensive programme published in *IOC Technical Series* No 7 (1970). The purpose of LEPOR is «to increase knowledge of the ocean, its contents and the contents of its subsoil, and its interfaces with the land, the atmosphere and the ocean floor and to improve understanding of processes operating in or affecting the marine environment, with the goal of enhanced utilization of the ocean and its resources for the benefit of mankind».

Compared with the purpose of IOC as given in its Statutes («to promote scientific investigation with a view to learning more about the nature and resources of the oceans through the concerted action of its members»), LEPOR contains an important new element, namely «*the goal of enhanced utilization of the ocean and its resources for the benefit of mankind*». The motivation for marine research is no longer purely scientific but clearly stresses the aspect of application of scientific results to the utilization of oceanic resources for the benefit of all nations. This new element necessarily affects the determination of IOC's goals, which have to be adjusted accordingly to include the utilization aspect and the needs of all nations. This implies that all countries should be entitled to an adequate share of those benefits. Since a high scientific and technological standard is required for making use of oceanic resources, the close interrelation of LEPOR to the field of technical assistance to developing countries is clearly paramount.

LEPOR itself is subdivided into two main parts :

– I. Scientific content of the expanded programme, and
– II. Practical problems of implementation.

In the scientific part there are enumerated and explained research projects for six main areas, namely :

1. Problems of ocean-atmosphere interaction, ocean circulation, variability and tsunamis ;
2. Living resources and their relations with the marine environment ;
3. Marine pollution ;
4. Geology, geophysics and mineral resources beneath the sea ;
5. The Integrated Global Ocean Station System (IGOSS) (Programme aspects) ; and
6. International investigations in specific regions.

Altogether some 50 research projects are described in the scientific part, which, although it reflects the broad spectrum of ocean research, is by no means exhaustive. Therefore, provision was made for criteria to be applied in the selection of further projects for inclusion in the expanded programme. Such projects

– must be characterized by broad and active participation of Member States ;
– must require international co-operative action ;
– must have a sound scientific basis ;
– must promise significant new information ;
– will contribute to enhanced utilization of the ocean and its resources ; and
– will help to meet the needs of developing countries.

There has been some criticism with regard to LEPOR, which some scientists considered to be more a «shopping list» than a programme. The field of ocean research is so wide and diversified, though, that many different avenues may lead to the goal of enhanced utilization of oceanic resources. A programme that contains many topics will certainly give some stimulation to all scientists and may be helpful to them when looking for financial support of their research.

The second part of LEPOR deals with the practical requirements necessary for the implementation of the programme. Herein actions are proposed regarding the following areas :

– Training, education and manpower requirements ;
– Data and information management ;
– Instrumentation and methods ;
– Technology and supporting facilities ;
– Supporting services ;
– Legal aspects of scientific investigation ;
– Integrated Global Ocean Station System (IGOSS) (Implementation aspects) ;
– Organization for implementation of LEPOR ; and
– Assistance to developing countries.

IOC Member States, when asked to express their interest and capabilities regarding the various projects contained in LEPOR, reacted by showing a very wide range of interest. The Commission subsequently charged its Group of Experts on Long-Term Scientific Policy and Planning (GELTSPAP) with the task of developing the scope and content of LEPOR by keeping under review relevant national and international programmes and by identifying fields of research requiring increased emphasis.

The group suceeded in transforming the expanded programme from a catalogue of separate projects into a coherent set of co-operative exercises and multinational experiments with corresponding oceanographic services and a strong component in training, education and mutual assistance.

Based on the GELTSPAP report, the IOC at its seventh session in 1971, after having adopted re-formulated criteria for assignment of priorities within LEPOR, designated the following eight research areas as *programmes of major importance within LEPOR*.

- Upwelling, including ocean atmosphere interaction.
- Survey of living resources.
- Coastal ecology and mariculture.
- Global investigation of pollution in the marine environment.
- Morphological charting of the sea floor.
- Systematic geological and geophysical surveys of continental margins, including the marginal seas.
- River discharge of sediments and along-shore transport.
- Physical research related to the Integrated Global Ocean Station System.

11.2 International Decade of Ocean Exploration (IDOE)

LEPOR is a long-term programme designed for implementation within several decades. The need was felt for an acceleration phase in which international co-operation in oceanography could be greatly increased within a comparatively short time span. The concept of an International Decade of Ocean Exploration (IDOE), introduced by the USA and welcomed by the United Nations General Assembly, was considered a useful initial element of LEPOR to provide such accelerative effect.

At its seventh session in 1971, IOC proposed «that IDOE should last from 1971 to 1980 and consist of appropriate national oceanographic activities of significant size and scope in which the participation of scientists from other nations is actively sought and achieved in the early stages of the programme».

Member States were encouraged to participate in IDOE and to submit to the IOC Secretariat detailed information on such programmes which they wished to be considered as components of IDOE. The criteria, which had to be fulfilled for such participation, were the following :

(1) the programme must be multinational ;
(2) the data collected during the programme must be submitted on a timely basis to the World Data Centre system ;
(3) the programme must serve exclusively peaceful purposes ;
(4) the programme must be scheduled to take place during the 1971-1980 period ;
(5) the programme must involve such scientific substance and international co-operation as to «accelerate» our knowledge and understanding of the ocean ; and
(6) participation of scientists from other nations must be actively sought and achieved in early stages of the programme.

The response of the Member States was remarkable. Quite a number of projects were submitted to be included into the framework of IDOE presented by IOC. These projects fell in the following four broad research areas :

- Environmental forecasting (16 programmes) ;
- Quality of the marine environment (5 programmes) ;
- Non-living resources of the sea floor (9 programmes) ; and
- Living resources of the sea, including the relationship between marine life and marine environment (1 programme).

A detailed account of the IDOE programmes has been published in *IOC Technical Series* No 13 (1974).

It should be noted that, with LEPOR, IOC has applied a different strategy in promoting co-operative research than it had done before LEPOR. In the past, IOC had identified a certain sea area (e.g. the Indian Ocean or the Caribbean) or a specific phenomenon (e.g. the Kuroshio) as in need of scientific investigation and had established co-ordinating machinery to be responsible for the implementation of the research programme. With IDOE this was different : IOC provided the general concept of increased international marine research within the decade 1971 to 1980, formulated relevant criteria and, for the rest, left the initiative for action to individual scientists, institutes, or countries who, in most cases, remained also responsible for the necessary co-ordination.

This new IOC strategy applied to IDOE leaves more freedom for ideas, initiative and action to the scientific community than the former co-operative investigations which were started, organized and co-ordinated by IOC. Perhaps this is the reason why IDOE had proved so successful.

There is, however, one mechanism newly introduced into IDOE by IOC, namely the *international workshop*. These are meetings organized and sponsored, often jointly with other interested agencies, by IOC in order to bring together working scientists with the task of :

- summarizing present knowledge and ongoing research projects relevant to a certain subject ;
- identifying major unsolved problems and need for further research work ; and
- making recommendations, when appropriate, for new scientific research and training programmes and co-operation.

These workshops are intended to promote co-operation at the scientist-to-scientist and institution-to-institution level. So far, 13 workshops have been held, i.e.

- 4 workshops on marine geology and geophysics at
  Bangkok, Thailand 1973 (regional)
  Kingston, Jamaica 1975 (regional)
  Suva, Fiji 1975 (regional)
  Mauritius 1976 (global)
- 5 workshops on marine pollution at
  Gaithersburg, USA 1974 (global)
  Monaco 1974 (regional)
  Penang, Malaysia 1976 (regional)
  Monaco 1976 (global)
  Port of Spain, Trinidad 1976 (regional)
  Santiago de Chile 1978 (regional)
- 1 workshop on the phenomenon known as 'El Niño' at
  Guayaquil, Ecuador 1974
- 1 workshop on ichthyoplankton at
  Mexico City, Mexico 1974
- 1 workshop on co-operative investigations in the north and central western Indian Ocean at
  Nairobi, Kenya 1976.

As the result of the Bangkok workshop, IOC, jointly with CCOP, established an IDOE co-ordinating mechanism, the *CCOP/IOC Joint Working Group on IDOE Studies on East Asia Tectonics and Resources (SEATAR)*. Similar developments are to be expected with regard to the studies on the phenomenon known as *'El Niño'* and to investigations on the *marine environment and resources in the north and central western Indian Ocean*.

11.3 Global Investigation of Pollution in the Marine Environment (GIPME)

The protection of the marine environment against pollution was early recognized to be an important issue. The intergovernmental conference on oceanographic research, held in Copenhagen in 1960, requested states to take, without delay, all steps in their power to prevent pollution of oceans and seas by radioactive and other harmful agents. Concrete action was started at the fourth session of the Commission in 1965 when the measures of other international organizations on preventing marine pollution were reviewed, with the aim of defining the role IOC might possibly play in the international effort to combat pollution of the ocean. Being aware of the great urgency of a better understanding of the processes governing marine pollution as well as the complexity of this subject, the Commission decided to establish an IOC *Working Group on Marine Pollution* with the task of studying how the Commission could further national and international endeavours in this important field. This group suceeded in working out, with the advice of SCOR and ACMRR, a clear definition of marine pollution and a classification of pollutants. It further stressed the necessity for better co-ordination on both national and international levels of marine pollution studies and preventive measures. Such co-ordination was finally achieved by the establishment of the Joint IMCO/FAO/Unesco/WMO/WHO/IAEA/UN/UNEP Group of Experts on the Scientific Aspects of Marine Pollution (GESAMP), with such terms of reference as to enable it to give pertinent advice also to the Commission. Consequently, the Commission, at its sixth session in 1969, decided to dissolve its Working Group on Marine Pollution which ended the first phase of IOC' s engagement in oceanographic aspects of marine pollution.

This subject was, however, taken up again in the Long-term and Expanded Programme of Oceanic exploration and Research (LEPOR) wherein the importance of oceanographic investigation of marine pollution is emphasized and a number of such studies is identified. These studies should enable IOC to review regularly the state of the ocean as regards pollution and to forecast long-term trends in order to assist governments in their endeavours of counteracting dangerous effects. For the purpose of LEPOR, marine pollution was defined as follows :
«Introduction by man, directly or indirectly, of substances or energy into the marine environment (including estuaries) resulting in such deleterious effects as harm of living resources, hazard to human health, hindrance to marine activities including fishing, impairing of quality for use of sea water and reduction of amenities».

In November 1970, the Group of Experts on Long-Term Scientific Policy and Planning (GELTSPAP), established by IOC to advise the Commission upon the implementation of LEPOR, proposed that a Global Investigation of Pollution in the Marine Environment (GIPME) be carried out within the framework of this long-term programme. The IOC Bureau and Consultative Council at its 12th session in March 1971, accepted this proposal and requested the Commission's scientific advisory bodies, together with GESAMP, to co-operate in developing the programme elements of such a global investigation. The advisory bodies SCOR, ACMRR, ACOMR and GESAMP responded promptly to this request and, based on their report conceived at Castellabate, Italy, the seventh session of IOC, in 1971, decided to establish the *Global Investigation of Pollution in the Marine Environment (GIPME)* as one of the major projects in the acceleration phase IDOE of LEPOR, to be organized by IOC in co-operation with the ICSPRO organizations and other appropriate bodies. The *International Co-ordination Group for GIPME* was subsequently created by the Executive Council of IOC at its first session in July 1972. It was composed jointly of scientists representing Member States or relevant organizations of the UN system who were actively involved in the conduct of marine pollution research programmes.

When taking such action, IOC was aware of the fact that additional funds for the IOC Secretariat and the scientific advisory bodies would be needed in carrying out the global investigation. The Commission therefore looked forward to getting financial support from the United Nations Conference on the Human Environment and, more specifically, from the emerging United Nations Environment Programme (UNEP), recognizing its obvious relevance to GIPME. UNEP has, in fact, greatly assisted financially in the planning and implementation of GIPME.

In spite of such favourably fulfilled prerequisites, the progress toward the development of a sound scientific strategy and a comprehensive implementation plan for GIPME was comparatively slow in the beginning. This was understandable in view of the complexity of the marine processes to be investigated. Nevertheless, after three meetings, the International Co-ordination Group for GIPME was able to submit to the Commission a *Comprehensive Plan for the Global Investigation of Pollution in the Marine Environment* which, after some amendment, was adopted by the Executive Council at its seventh session in 1976. The Council further decided to combine the Comprehensive Plan with a report prepared by the IOC/ICES Working Group on Baseline Study Guidelines and to publish these documents together in the *IOC Technical Series* (No 14, 1976), in order to provide guidance to Member States wishing to participate in the global investigation.

Parallel to this action, the International Co-ordination Group for GIPME recommended in 1973 that IOC should take the initiative in preparing a preliminary report on the *Health of the Ocean* which was intended to serve as a starting point for a continuing review of the state of the ocean with regard to pollution to be carried out by IOC. This report was written by a marine chemist (Dr. E.D. Goldberg, USA) and published by the Unesco Press on behalf of IOC in 1976. It deals with five kinds of major marine pollutants :

- halogenated hydrocarbons ;
- radioactivity ;
- heavy metals ;
- petroleum hydrocarbons ; and
- litter.

For each of these all available information on their impact on the marine environment (air, water, organisms, sediments) resulting from man's activities as compared to natural influences has been assembled. The main difficulty lies in the fact that most of the pollutants already identified are in extremely low concentrations, both in sea water and in marine organisms, so that measurements have to be made near the limit of sensitivity of the analytical method used hereto for. In spite of these low concentrations, concern about marine pollution is justified, because so very little is known about what happens to pollutants once they have entered the ocean. They may be diluted, stored, altered, decomposed to harmless substances or converted into a more toxic form, taken up and accumulated by organisms, passed into the sediments or be transferred to the atmosphere. There is always an imminent serious danger of deleterious effects on the marine environment, on the living and non-living resources of the sea and thus on man.

In view of these difficulties and uncertainties, it is very important that the Comprehensive Plan for GIPME provides an international scientific framework which will help to co-ordinate national and regional efforts and to concentrate them on the ultimate objective of GIPME, namely the establishment of a sound scientific basis for the assessment and regulation of the marine pollution problem.

First priority in GIPME has been given to baseline studies, conducted on a national or regional scheme and aimed at building up information on the global distribution of marine pollutants which will serve as a basis for future statements on the health of the oceans. Equal weight is attached to research activities dealing with inputs, pathways, sinks, effects, and dose/response relationships of major pollutants in the ocean. Further, calculations of mass balance and studies of transfer processes between major reservoirs (ocean, atmosphere, organisms, sediment) are needed. For any regulatory action regarding marine pollution to be efficacious, it is necessary to know the degree of exposure acceptable for man or marine organisms. To achieve this, basic standards for exposure to pollutants either for man or for marine organisms have to be determined. Since physical, chemical, biological and geological problems have to be tackled and solved, GIPME really is a prototype of an interdisciplinary exercise.

In addition to the scientific framework outlined above, the Comprehensive Plan for GIPME contains recommendations on how this plan should be implemented. Regional marine pollution workshops, intended to design and co-ordinate the implementation of regional baseline studies, are considered to be an important first step, as mentioned in Section 11.2 above.

After the planning phase of GIPME had ended, the ninth session of the IOC Assembly in 1975 disbanded the International Co-ordination Group for GIPME and replaced it by a *Working Committee for GIPME*, with the responsibility for promoting the implementation of the Comprehensive Plan for GIPME. Unlike the International Co-ordination Group, which developed the plan and had only a limited membership, the Working Committee, besides its somewhat greater autonomy, is open to all IOC Member States. In fact, the co-operation and participation of a great number of them are essential if GIPME is to reach its goal.

The Working Committee for GIPME started its work in 1976, reviewed regional components of GIPME and identified specific objectives and priorities for regional marine pollution research projects, including uniform formats for reporting on such activities. The Committee is concerned with selecting standard methods of sampling and analysis as well as with the development of very careful intercalibration procedures which are needed for carrying out effective baseline or monitoring programmes. The programmes of monitoring marine pollution in coastal waters, as well background levels of selected pollutants in open-ocean waters, were given particular attention.

In the beginning, participation of IOC Member States in the Working Committee for GIPME was somewhat slow. Gradually, though, awareness has been growing that marine pollution is a global problem which requires the full co-operation of all maritime countries.

An overview of the main aspects of global marine pollution has been published in *IOC Technical Series* No 18 (1977).

12 Oceanographic programmes associated with atmospheric research

Oceanographers are well aware of the intimate relations between the atmosphere and the ocean. Over large seas, the atmosphere receives energy and matter from the ocean through the transport of sensible and latent heat as well as of substances. Another very prominent interaction is caused by the wind which is the driving force of ocean surface waves and currents. Ocean and atmosphere form an interacting and feed-back system on a very large scale. This fact is reflected in many of IOC's earlier statements and resolutions, including the programme of the International Indian Ocean Expedition and the General Scientific Framework for World Ocean Study.

A formalized approach to this subject took place in 1965 when the fourth session of the IOC decided to establish an *IOC Working Group on Ocean-Atmosphere Interaction* to consider the operational aspects and opportunities for intergovernmental action in this field. Since this group arrived at the conclusion that a Joint IOC/WMO Working Group on Ocean-Atmosphere Interaction would be a much more appropriate body to deal with the subject, the IOC group was dissolved by the fifth session of the Commission in 1967, in order to negociate with WMO the establishment of such a joint group. WMO, however, trying to avoid duplication of its effort, felt that a joint WMO/IOC/ICSU Panel on Ocean-Atmosphere Interaction, to be established within the framework of the *Global Atmospheric Research Programme (GARP)* of WMO and ICSU, would offer a better solution. GARP is a programme directed to studying those physical processes in the atmosphere, the understanding of which is essential for increasing the accuracy of weather forecasting over periods from one day to several weeks and for determining the physical basis of climate. Thus, IOC's interest and engagement in ocean-atmosphere interchange, as far as the research side was concerned, became associated with GARP, while governmental operational and planning aspects of ocean-atmosphere interaction were dealt with by the IOC Working Committee for IGOSS (see Section 14.2).

This was in conformity with recommendations by GELTSPAP which stated that oceanographers must take every opportunity for conducting oceanographic investigations in conjunction with relevant meteorological experiments under GARP, both to strengthen these experiments and to utilize the facilities and the information on atmospheric input thus provided for oceanographic research purposes.

The first concrete IOC action with respect to GARP was the arrangement for oceanographic participation in the *GARP Atlantic Tropical Experiment (GATE)*.

The objective of GATE was to study the processes of energy transfer from the ocean to the atmosphere and within the atmosphere, taking place in the Atlantic tropical region in various scales of motion. In particular, the aim of the activity was to investigate the internal structure of cloud clusters and to estimate the vertical and horizontal transport of heat and moisture, as well as of momentum associated with such systems, and to relate them to atmospheric motions of larger scales.

The IOC Bureau and Consultative Council, at its 12th session in 1971, requested SCOR to identify the oceanographic processes that could be studied in conjunction with GATE. Further, it was thought necessary that an oceanographer should work with the International Scientific and Management Group (ISMG) for GATE in order to ensure adequate co-ordination of oceanographic projects with the meteorological experiment.

SCOR, realizing that GATE provided a unique opportunity to investigate the response of the ocean to atmospheric forcing on various scales, responded to this request and developed an oceanographic programme for GATE composed of three experiments related to different scales.

(1) *C-scale experiment*

Investigation of physical processes in the top layer of the ocean (0 - 200 m) disturbed by local atmospheric forcing within a 100 km triangle, including :

- surface waves ;
- internal waves ;
- mixed layer development ; and
- fronts in the thermocline.

(2) *B-area experiment*

Investigation of the response of the ocean (0 - 1000 m) to atmospheric forcing on the B-array scale (up to 1000 km), including :

- mixed layer budget study ; and
- main thermocline response experiment.

(3) *Equatorial experiment*

Study of the equatorial current system and its relationship to the atmosphere, including :

- physical oceanographic description of the equatorial current system ;
- dynamics of the Equatorial Undercurrent ; and
- spatial and temporal scales of the Equatorial Undercurrent.

The GATE programme was successfully carried out in the summer and autumn of 1974. About 40 research vessels participated, in addition a great number of buoys, moorings and aircraft were deployed. The oceanographic data were collected and processed at a special Oceanographic Sub-programme Data Centre for GATE in Brest, France. Scientific results of GATE were presented and discussed at several workshops and at the GATE Symposium on Oceanography and Surface Layer Meteorology, 1978, in Kiel, Federal Republic of Germany, at which time the oceanographic component of GATE was terminated. Arrangements were made for the publication of a synoptic atlas, composed of 3 volumes, to present the oceanographic data collected during GATE in a form convenient for use by oceanographers who do not have access to the archived data in computer format or who require a preliminary survey of the material available before undertaking a detailed investigation of part of the archived data.

From the very beginning of GARP it was recognized that a period would be needed during which the entire global atmosphere was observed. This period is called the Observational Phase of the *First GARP Global Experiment (FGGE)* which, after a preparatory phase during 1978, started in December 1978 and will last until November 1979. It will include two Special Observing Periods (SOPs), characterized by a major observational effort, during January-February and May-June 1979. The Observational Phase will be followed by a Research and Evaluation Phase extending well into the 1980s.

This global meteorological experiment again provides an opportunity for oceanographers to study, in various areas, ocean processes originating from atmospheric forcing. On the other hand, meteorologists are strongly interested in the participation of oceanographic research vessels in FGGE because of the practical advantage of such an enterprise ; about 50 ships are urgently needed to serve as platforms for upper air ascents in the tropical oceans. Therefore, the IOC Executive Council, at its second session in 1973, requested SCOR to propose the elements of an oceanographic programme associated with FGGE. SCOR reacted favourably and, in 1977, presented a report containing details of the *FGGE Tropical Oceanographic Programme* concerned with the global study of the dynamics of the equatorial current system, including :

- mass and heat transport ;
- low frequency variability (3 to 50 days) ;
- atmospheric forcing ;
- interactions with extra-tropical flows ; and
- mixing, upwelling and productivity.

Evidence of all of these phenomena has been verified in the tropical regions of the Atlantic, Indian and Pacific Oceans although it has been found that they occur with varying degrees of similarity in each of these oceans. Consequently, different approaches had to be planned for the research. The IOC Assembly, at its tenth session in 1977, acknowledged that FGGE is a global programme of major significance to oceanographic science, which, during the next two years, should be given high priority in the programmes of the Commission. IOC's role in implementing this programmme will lie in :

- handling all FGGE oceanographic data management activities ;
- facilitating the entry and passage of ships participating in FGGE into its Member States' ports and waters under national jurisdiction ;
- developing, promoting and further strengthening IGOSS to assist in data collection ;
- being responsible for the operational and co-ordination aspects of FGGE oceanographic programmes.

The IOC Assembly, at its ninth session in 1975, established, together with SCOR, an IOC/SCOR *ad hoc* Task Team to identify a *Comprehensive Oceanographic Programme related to GARP* and prepare a plan consisting of present and possible future IOC activities which could benefit from and provide support to GARP. This plan was to embrace not only the oceanographic programme for FGGE but also the oceanographic component related to the so-called second objective of GARP, namely to work towards an improved understanding of the physical basis of climate. Although the planning for a global programme of climate research is still in the formative stage, it is clear that, in addition to atmospheric processes, the study of ocean processes will have to be fully integrated into such a programme. It is assumed that understanding of climate variability can best be gained by approaching the problem both from diagnostic, observational grounds as well as from theoretical modelling points of view. Within such a framework, a set of models comprising the whole atmosphere-ocean-cryosphere system will have to be developed and verified.

The Comprehensive Oceanographic Programme related to GARP, as designed by the IOC/SCOR Task Team, was approved by the IOC Executive Council at its eighth session in 1977 and published in the *IOC Technical Series* (No 17). Thus, it can be used as guidance for the Commission's further activities related to GARP. With regard to the second objective of GARP, much further study is necessary.

CHAPTER IV

Services and assistance

13 Training, Education and Mutual Assistance in the marine sciences (TEMA)

13.1 Initial stages

From the very beginning of IOC's existence the training and education aspect has received considerable attention. This is understandable with a body like IOC, established to promote international co-operation and composed of Member States in very different stages of development in the field of marine science. Improving the marine scientific capabilities of the less developed states is an absolute necessity if sufficient participation in IOC's programme is to be acheived.

The training aspect was first reflected in the idea of an *international research and training vessel* to be operated by Unesco. However the IOC, at its first session in 1961, realized the difficulties inherent in implementing such a proposal, and found that shipboard training could be carried out more efficiently and less expensively aboard national research vessels. So the idea was not developed any further at this time. Instead, the nations which operated national oceanographic research vessels were encouraged to accept scientists from other nations for training.

A comprehensive discussion on TEMA aspects took place at the third session of IOC in 1964 under the heading «Means by which the Commission can assist its Members in the development of national oceanographic programmes». Herein the following measures were suggested :

- Each country should examine its position in oceanography with respect to several categories as e.g. manpower, facilities (ships, laboratories, equipment, libraries), present efforts and interests, and future needs, and prepare a relevant report for consideration by its fellow Members of the Commission.
- Visiting committees should be appointed by IOC upon the request of a Member State to study, review, or advise on a national marine programme.
- Encouragement should be given to sister relationships between oceanographic institutions of advanced and developing countries.
- Information on the availability of inexpensive, simply-operated instruments and on means of distributing them should be provided by the IOC Secretariat.

The functions of IOC were considered to be those of a catalyst in stimulating oceanographic activities in its Member States through its own programmes and with Unesco's assistance.

As a consequence of this discussion, in 1964, IOC decided to establish a *Working Group on Mutual Assistance* to carry out the tasks mentioned above and, further, to help Member States in obtaining the financial and technical assistance needed for the development of marine sciences. In addition, advice on curricula and methods for educating marine scientists and technicians as well as arrangements for shipboard fellowships on research vessels for these persons, should be provided.

Although the decisions thus taken appeared to be reasonable and useful, progress in this field turned out to be comparatively slow. The national reports did not come in quickly and completely from all Member States, nor did the IOC Bureau succeed in forming the Working Group on Mutual Assistance promptly but instead decided to act itself as that working group, although it was clear that the Bureau alone could not adequately implement all parts of the work assigned to the group. From the further discussion of this matter, it could be taken that the global approach by IOC in the field of mutual assistance was not appropriate due to the diversity of needs in the various regions, and that a regional approach would promise better results.

Nevertheless, the IOC, at its fourth session in 1965, recognizing the urgent need of action in this field reaffirmed the decision taken at its third session, concerning the establishment of the Working Group on Mutual Assistance and re-defined its terms of reference. Finally, this group started work ; it met twice in 1966/67 and submitted a number of recommendations to the fifth session of IOC in 1967, which introduced several new aspects into the programme :

- it was recommended that Member States should create national oceanographic committees (where these did not already exist) to serve as national focal points for IOC activities;
- the Commission recognized the need for establishing close contacts between Member States in the various regions, in order to strengthen regional activities in oceanography;
- the Commission recognized that mutual assistance between Member States must and can be supplemented by technical aid provided by international agencies as Unesco and UNDP;
- the Commission recognized the extreme importance of training and education for the further development of oceanography and the urgency of better adjustment of training activities to the particular specific needs of its Member States in various regions. This was also a reaction on Resolution 2172 (XXI) of the General Assembly of the United Nations requesting the strengthening of marine education and training programmes.

As a result of these deliberations, the IOC, at its fifth session in 1967, decided to establish a Working Group on Training and Education in Oceanography in addition to the existing Working Group on Mutual Assistance which was asked to continue its work. This clearly shows the great importance IOC wished to attach to the matters of training, education and mutual assistance although there was some danger of overlapping of work and doubling of effort between the two groups concerned. Tendencies for combining these two groups were, therefore, already recognizable in 1969 after the two groups had held separate meetings. As an intermediate measure it was resolved that the two groups should continue their work separately for the time being but, if they both held meetings, they should meet at the same time and place, in order to facilitate discussions.

This comparatively modest progress was, in a way, made up for by other measures, for instance, by Unesco's successful Fellowship Programme in Marine Science and by the establishment of a post in Unesco's Office of Oceanography for a Training and Education Officer. The IOC Secretariat was charged with maintaining an up-to-date information file on training opportunities.

Eventually, the need for these important activities to be a part of IOC's programme was recognized in connexion with the Long-term and Expanded Programme of Oceanic exploration and Research (LEPOR), and as a result greater emphasis was placed on them. Although the joint meeting of the two groups mentioned above, held in Malta in January 1971, did not attract great participation of Member States, it was able to produce a comprehensive set of proposals which proved that programmes of training, education and mutual assistance are primary functions of the Commission and vital to the developing countries. IOC, recognizing that it had not yet succeeded in ensuring full participation of all its Member States in its activities, although such full participation would be required for the implementation of LEPOR, felt the urgent need of attaching more importance to the aspects of training, education and mutual assistance. At its seventh session in 1971, for the first time in IOC's history, a special session committee was charged with dealing with those questions, particularly with the recommendations of the joint Malta meeting. The decisions which resulted can be summarized as follows :

- A well-balanced programme for conducting training courses at different levels should be developed.
- Existing programmes for teaching marine science at different levels (including curricula, texts, and reference materials) should be compiled.
- Recommendations as to the introduction of marine science elements into the curricula of primary and secondary schools should be developed and communicated to Member States.
- Member States were urged to designate national contacts for subjects related to education and training in marine science.
- Up-to-date directories of training facilities and opportunities, including shipboard fellowships, should be published regularly.
- Selected textbooks and manuals in marine science should be translated into languages of developing countries where the need is most urgent.
- Financial support for these programmes should be sought from Unesco and UNDP, or be provided by Member States' contributions to the IOC Trust Fund in the use of which highest priority was given to education and training in marine science.

Subsequently the organizational structure of IOC's activities in this field was streamlined in so far as, in 1972, the two Working Groups on Mutual Assistance and on Training and Education were replaced by a single working group to deal with these tasks. At the eighth session of the IOC Assembly in 1973, this working group was renamed and became the *Working Committee for Training, Education and Mutual Assistance* (TEMA).

13.2 IOC's Working Committee for TEMA

With the establishment of the Working Committee for TEMA a new era began in IOC's activities in the field of training, education and mutual assistance. Based on the concept that TEMA should become the focal point of all IOC activities until a higher scientific and technological standard is achieved in all Member States, IOC developed a new and more efficient strategy in this important field. The urgent need to incorporate TEMA components in all future scientific programmes of the Commission was recognized. This decisive new approach was based on the recommendations of the first session of the Working Committee for TEMA held in 1973 and transmitted the same year to the eighth session of the IOC Assembly. At this time a wide spectrum of TEMA Activities, comprising the following themes, was launched :

- Training of marine science administrators ;
- Participation in oceanographic cruises ;
- Translations of important marine science texts ;
- Assessment of training needs by convening small groups of experts on a regional basis ;
- Use of the IOC Trust Fund for TEMA purposes ;
- Regional training centres ;
- Annotated bibliography of marine science textbooks, reference books and manuals ;
- Directories of experts in the marine sciences ;
- Participation of young foreign scientists in research and training projects of developing countries ;
- Programme of work and budget for TEMA.

In order to carry out these tasks, an additional staff member was recruited for the IOC Secretariat who was solely responsible for TEMA matters.

Among the activities mentioned above the assessment of the training needs of developing countries at regional *ad hoc* TEMA meetings proved to be particularly effective and helpful. This is particularly true because such needs are very diverse and can be defined much better for certain limited regions rather than be approached globally. So far, seven such meetings have been held :

- Mexico City, Mexico, April 1975,
- Casablanca, Morocco, June 1975,
- Manila, Philippines, September 1975,
- Cairo, Egypt, January 1976,
- Montevideo, Uruguay, November 1976,
- Karachi, Pakistan, March 1978.

The Working Committee for TEMA, at its second session at the UN Headquarters in New York, 1977, evaluated the results so far obtained by this action. Subsequently, the IOC Assembly at its tenth session later that year was able to base its decisions on the recommendations of this session of the Working Committee and, further, had at its disposal an informative and critical study on possibilities and alternatives of action in IOC regarding TEMA activities prepared by a working group in Mexico. The Assembly redrafted the terms of reference of the Working Committee for TEMA which, in addition to its tasks of reviewing the needs of Member States regarding training, education and mutual assistance and of recommending corresponding TEMA programmes, was, «in consultation with other subsidiary bodies of IOC and with the ICSPRO agencies, (to) *arrange* the provision of scientific and technical education, and the transfer of relevant technology and technical assistance to developing Member States in marine science aspects of ocean affairs in order to build up their national capabilities to participate fully in ocean research of interest to them, including IOC programmes, and to achieve self-reliance in marine sciences as a whole».

According to its Statutes, the Commission is not regarded as an operational organization in the field of training and education, but as a promotional, co-ordinating and advisory agency, while the relevant operational functions are vested in the ICSPRO agencies, particularly in Unesco. It is in the light of these general rules that the functions of the Working Committee for TEMA must be seen.

13.3 IOC's Voluntary Assistance Programme (IOC-VAP)

At its ninth session in 1975, the IOC Assembly instructed the Secretary to develop a plan for an IOC Voluntary Assistance Programme similar to that established and successfully executed by the World Meteorological Organization.

A relevant draft, examined and amended by the second session of the Working Committee for TEMA, was before the IOC Assembly in 1977. After some discussion, this document was accepted as the basis for a Voluntary Assistance Programme within the Commission (IOC-VAP), and the rules for the utilization of this programme were established.

This programme has introduced a mechanism for technical assistance in marine science under the co-ordination and responsibility of IOC and with the active participation of the ICSPRO agencies concerned. IOC-VAP constitutes an important supplementary source of support in this field, in addition to national programmes as well as to bilateral or multilateral programmes of assistance in marine science from ICSPRO and other agencies.

IOC-VAP is comprised of the following components :

- programmes for training, education and mutual assistance, including TEMA components of ongoing IOC programmes ;
- the strengthening of educational and research institutions in developing countries ; and
- any other project that the IOC and the donor nation deem to be beneficial for the development of marine sciences and the proper usage and protection of the marine environment.

The IOC-VAP is maintained by voluntary contributions received from IOC Members States for the purpose of meeting official requests from other IOC Member States which have been approved by the Executive Council, based on the special criteria for the utilization of this programme. Such contributions may take the form of money, equipment or services. It is hoped that the IOC-VAP will bring about a decisive step forward in the development of oceanography all over the world.

14 Ocean services of IOC

It is the characteristic of such a «service» that specified information is provided to interested users in a more or less regular fashion or upon special request or whenever such information is needed to avoid damage and destruction. Service aspects were already recognizable at the beginning of IOC's activities, evidenced by proposals and discussions made which dealt with the exchange of oceanographic data, the establishment of a network of fixed oceanographic stations and with the creation of an international tsunami warning system.

The objectives of such ocean services are twofold in that services may be rendered :

- to the scientific community, and/or
- to the governments and the public.

An example of the first form of ocean services is the collection, archiving, storage, retrieval and exchange of raw and processed ocean data which are needed for marine research purposes.

Under the second form of ocean services we may consider the supply of oceanographic information of such kind that governments and the public need in order to be able to «manage» ocean affairs, such as to :

- minimize the hazards of ocean and atmosphere ;
- make proper use of oceanic living and non-living resources ;
- safeguard the marine environment ;
- improve weather forecasting and marine transport ;
- protect and develop coastal areas.

The various services offered by the Commission will be described below.

14.1 International Oceanographic Data Exchange (IODE)

Two main parts of IOC's system of oceanographic data exchange were not created by the Commission itself but were already in existence when IOC came into being in 1960. These are the *World Data Centres 'A'* (in Washington, D.C., USA) *and 'B'* (in Moscow, USSR), financed by the United States and the Soviet Union. Both centres had been established for collecting geophysical data of various kinds, including oceanographic data, during the International Geophysical Year (IGY), 1957/58. This arrangement was to facilitate access to oceanographic data on a world-wide scale. Instead of having to address themselves to various national organizations, scientists can request and receive data necessary for their studies directly from the World Data Centres (WDCs), which are prepared to supply copies of material in the data centre to any scientific body or investigator in any country at a cost not exceeding the cost of copying and postage. Through this new form of data dissemination, it became possible to investigate oceanic phenomena on a global scale.

Apart from these two World Data Centres, there exist a number of specialized data centres, which are either disciplinary or regional in character. *The Permanent Service for Mean Sea Level* in Birkenhead, England and the *International Hydrographic Bureau* in Monte Carlo, Monaco, charged with collecting mean sea level and bathymetric data, respectively, as well as the *FAO Fishery Data Centre* in Rome, Italy, are of disciplinary character. The *International Council for the Exploration of the Sea* in Charlottenlund, Denmark, engaged in collecting oceanographic data from the North-east Atlantic Ocean and its marginal seas, has operated a regional centre for many years.

IOC, recognizing that full and expeditious exchange of oceanographic data forms the core of successful scientific co-operation in the marine field, decided at its first session in 1961 that the system of World Data Centres, as established during the International Geophysical Year, should be continued. It also recommended that all oceanographic data obtained by ships and automatic data platforms outside territorial waters, as part of a «declared national programme» or an international co-operative oceanographic expedition should be exchanged, as from 1 January 1960. «*Declared National Programmes*» (DNPs) are national oceanographic activities of IOC Member States which are declared as being carried out with the intention to exchange internationally the data resulting therefrom.

Further, the Commission recommended to its Member States either the establishment of «*National Oceanographic Data Centres*» (NODCs) or the nomination of appropriate «*Designated National Agencies*» (DNAs), in order to facilitate the collection, processing, analysis, and exchange of oceanographic data.

With the task to develop further and improve the system of oceanographic data management, a working group of experts was formed, composed of representatives from World Data Centres 'A' and 'B', from the Permanent Service for Mean Sea Level and from relevant international organizations such as the International Council for the Exploration of the Sea (ICES), the International Hydrographic Bureau (IHB) and the World Meteorological Organization (WMO).

Regarding the general principles and details of data exchange, the IGY Data Centre Manual was used as the basic document. It was developed by the working group into a provisional guide for the exchange of oceanographic data, which was approved as such by the second session of the Commission in 1962. A new text of this guide was subsequently prepared by the working group, adopted by the third session of IOC in 1964 and published as *IOC Technical Series* No 1 (1965) under the title «Manual on international oceanographic data exchange».

In spite of these endeavours to organize and facilitate the submission of oceanographic data to the WDCs, the inflow of such data progressed very slowly. In particular, there was a certain reluctance on the side of the Member States to indicate which of their national marine research activities were «Declared National Programmes» with the obligation of submitting the resulting data to the World Data Centres. Further, the use of several different formats, i. e. the lack of a uniform standard format, presented an obstacle to data exchange. In addition the need was also felt to extend the types of data being exchanged, so far more or less limited to physical oceanography and bathymetry, to other areas of marine science, such as chemical, biological, geological and marine pollution data.

The Bureau and Consultative Council, at their fifth meeting in 1965, as well as the Commission itself at its fourth session in 1965, discussed these difficulties in detail and asked the *Working Group on Data Exchange* to pay particular attention to the introduction of methods whereby the flow of data to the World Data Centres could be increased and speeded up.

In the years to follow, the situation in data collecting has improved considerably. In 1968, more data were transmitted to the World Data Centre System than during the first seven years of its existence. This was partially due to the submission of data hitherto unavailable to the WDC System, partly because more and more IOC Member States had established National Oceanographic Data Centres (NODCs) which form a necessary link between the national oceanographic laboratories and the World Data Centres (Oceanography). In order to provide the necessary guidance to this enhanced activity, a revised version of the «Manual on international oceanographic data exchange» was published in 1967.

New problems then arose in connexion with the rapidly-growing amount of data from continuously-recording and remote sensors and automatic logging devices as well as with the data archiving and exchange requirements related to the new *Integrated Global Ocean Station System (IGOSS)* producing real-time and near-real-time data, i.e. transmitting oceanographic data at or near the time of observation.

In addition, scientific advisory bodies of the Commission, like those present at the Helio Cabala Meeting (see section 7), pointed to possible inadequacies in the existing data exchange system in coping with the flood of diversified data resulting from expanded programmes in marine science. The need for improvement in this respect became particularly urgent when, in 1969, the *Long-term and Expanded Programme of Oceanic exploration and Research* (LEPOR) was adopted which, in its section on implementation, enumerated the following problems of data and information management in need of solutions :

- improvement and consolidation of bibliographic and related information services ;
- early exchange of plans and preliminary results of observational programmes ;
- integration of real-time exchange of oceanographic data with the meteorological system ;
- development of methods for storage and retrieval of biological, geological and geophysical data ;
- automation of international data banks and improved programmes and methods for making their contents available ;
- development of standardized and/or computer-compatible data formats ;
- timely establishment or improvement of international inventories of ocean data and samples and provision for centralized cataloguing of sea data available from various private and public sources ;
- strengthening of the system of sorting centres for biological material.

There followed a very active period of IOC's *Working Group on International Oceanographic Data Exchange* (IODE), now composed of representatives of all interested Member States, which worked effectively with various task teams and *ad hoc* groups carrying out their duties by correspondence. They considered the interdisciplinary and interorganizational implications inherent in data and information management and referral, in such areas as air-sea interaction, marine resources, marine pollution, and recreation and safety.

As a result, a number of proposals were submitted to the seventh session of the IOC in 1971 which took relevant decisions with a view to improving and increasing the effectiveness of the system for the international exchange of oceanographic data in a multi-disciplinary fashion. Among those decisions was the adoption of an interim standard international data inventory form ROSCOP (Report of Observations/Samples Collected by Oceanographic Programmes), for the timely submission of information on expeditions carried out in accordance with Declared National Programmes and programmes of international co-operative research. This new marine data inventory (ROSCOP) programme includes information on chemical, biological, geological, geophysical and pollution data. ROSCOP is intended to fill the gap between the first announcement of an oceanographic programme and the eventual cataloguing of data actually received by the WDCs.

Further, the Working Group on International Oceanographic Data Exchange was asked to prepare, as expeditiously as possible, compatible formats for the international exchange of oceanographic data and to concentrate on the preparation of standard forms for conventional data.

The new regulations required the issue of a third, revised edition of the «Manual on International Oceanographic Data Exchange» which was published as *IOC Technical Series* No 9 in 1973 and is subject to continual revision and up-dating following the progress of data exchange procedures. Subsequently, the fourth edition appeared in 1976 as No 9 of the series *IOC Manuals and Guides*. In order to cover the IGOSS aspects a «Manual on IGOSS Data Archiving and Exchange» was also conceived and published.

The main features of the data exchange procedures, as identified in the manual mentioned above, are the following :

- *Standard observations* are environmental observations or measurements made from research vessels, other ocean platforms with generally accepted types of instruments and methods widely known. Such data, when submitted for general use, either require no correction, or such corrections are generally known and available. Data resulting from such observations or measurements are exchanged through the World Data Centre System.
- *Non-standard, experimental or other special observations* are generally to be retained by the originating countries and exchanged only upon request.

The following procedures apply to the *submission of data, information and inventories* :

- Advance information on oceanographic programmes sent to the IOC Secretariat for dissemination as Information documents.
- Inventory form ROSCOP, confirming that announced programmes have been carried out, sent to NODCs and WDCs.
- Individual data sets, preferably in forms facilitating machine processing, forwarded to the WDCs or the relevant disciplinary data centre via the relevant NODC as expeditiously as possible after the completion of the programme, i.e. within one to two years, depending on the type of data.
- World Data Centres are responsible for the provision of data and information to any qualified requester in the scientific community. They publish, on a periodic basis, inventories of their data holdings.

The Commission further encouraged National Oceanographic Data Centres (NODCs), especially those with facilities for automation, to assume the function of regional data centres wherever feasible and to perform data processing and analysis services on a voluntary basis for data collected by countries lacking those facilities, thus providing mutual assistance in the international exchange of automated data.

At the eighth session of its Assembly in 1973, the Commission decided on measures for rationalizing the structure of its subsidiary bodies and, in the course of this action, the *Working Group on International Oceanographic Data Exchange* (IODE) was renamed *Working Committee*. The terms of reference were expanded to include matters relating to information management as well as data exchange, with somewhat greater autonomy given to the Committee. «Information», for these purposes, was defined as published and unpublished literature, formal, semi-formal and informal written and oral communications.

By 1973, 34 IOC Member States (45% out of the total membership of the Commission at that time) had established National Oceanographic Data Centres or designated National Agencies for data exchange purposes. In order to assist other Member States in building up suitable national structures for data exchange in the marine field, the IOC Working Committee on International Oceanographic Data Exchange conceived a «Guide for Establishing a National Oceanographic Data Centre» which was published in the series *IOC Manuals and Guides*. The present extent of the International Oceanographic Data Exchange System is shown in Fig. 11. To date, 75 Member States (77%) have provided input into the world oceanographic data banks.

In the years from 1973 to 1977 the deliberations of the Working Committee on IODE aimed at further improvement and advancement of the data exchange system. The following actions were taken :

- A second edition of ROSCOP was prepared and distributed to Member States.
- A second level inventory form ROMBI (Results of Marine Biological Investigations), developed by SCOR and ACMRR, was tested to facilitate dissemination and exchange of marine biological data.
- The use of the International Geological/Geophysical Cruise Inventory (IG/GCI) form proposed by IUGS/CMG was also endorsed as an aid in exchange of geological and geophysical data through WDCs.
- New data formats for international exchange of marine geological data were introduced on an experimental basis.
- The IOC General Format for exchange of oceanographic data (now GF-3) was recommended for use by Member States and tested.
- Consideration is being given to the ways in which satellite and airborne sensed oceanographic data can be incorporated into the archiving and exchange system.
- The concept of Responsible National Oceanographic Data Centres (RNODCs) was investigated, and a pilot programme leading to the establishment of RNODCs was approved. Their primary function is to aid the WDCs by providing various types of services (statistical data summaries, data displays, etc.) and to assist NODCs and DNAs that request such help with the conversion of data for subsequent submission to the WDCs.
- For either specific projects, specific regions or specialized types of data, IOC co-operated with relevant international organizations (WMO, FAO, ICES, IHO, UNEP) in the field of interdisciplinary and inter-organizational data and information management and referral and contributed to the development of a *Marine Environmental Data Information Referral System* (MEDI). Its objective is to provide information on the location,

Figure 11 - **International Oceanographic Data Exchange System** (from IOC Biennial Report for 1976-77)

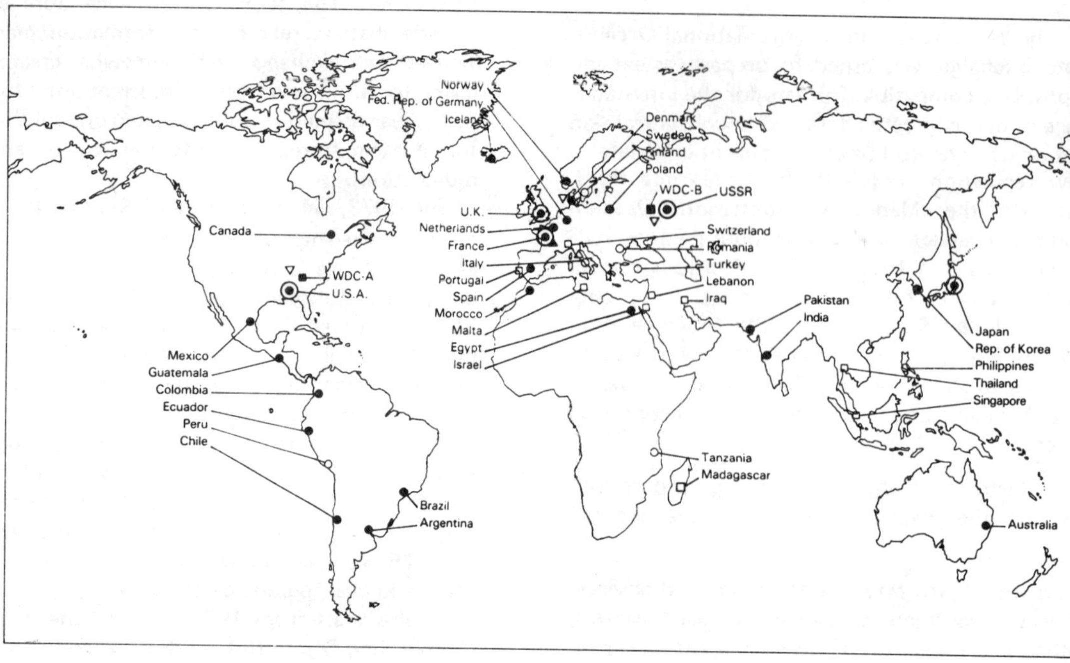

△ Specialized Centres
International Hydrographic Organization (IHO), Monaco
The Permanent Service for Mean Sea Level (PSMSL), U.K.
FAO Fishery Data Centre - FAO, Italy

● National Oceanographic Data Centre (NODC)

○ Designated National Agency (DNA)

□ National Co-ordinator for IODE

▲ MEDI Co-ordination Centre (IOC/Unesco, France)

■ World Data Centres -A and -B (Oceanography)

▽ Regional Data Centres
IOCARIBE - USA
CIM - USSR
ICES - Service Hydrographique

◉ Responsible National Oceancgraphic Data Centres (RNODCs)
IGOSS BATHY/TESAC - USSR, USA, France, Japan
IGOSS MAPMOPP - USA, Japan
Wave Data - U.K.
WESTPAC - Japan
FOY - USA

availability and characteristics of marine environmental data. The co-ordinating function for MEDI has been assigned to the IOC Secretariat.

– IOC further co-operated with FAO in setting up an *Aquatic Sciences and Fisheries Information System* (ASFIS) providing scientific and technical information in the fields of marine and freshwater science and technology with main emphasis on fisheries aspects.

Through these endeavours, the amount of oceanographic data stored in the World Data Centres grew steadily. The WDCs for oceanography presently complete statistics of data exchange in annual and semi-annual reports. By December 1978, more than 1.2 million observations have been exchanged in the WDC system. Figure 12 A summarizes annual data receipts for all types of data ; Figure 12 B that for oceanographic serial stations. If those data were presented according to their geographical positions, the picture would reveal how uneven the coverage of the oceans with such observations really is. Apart from some coastal regions and marginal seas, mostly in the Northern Hemisphere, where the coverage appears to be fair, there are vast ocean areas, particularly in the Southern Hemisphere, where the network of measurements is extremely sparse. This is all the more true if one realizes that those measurements refer to the ocean water and not to the ocean floor. Consequently, they have only a limited representation in time, because the oceans are in permanent motion and characterized by a substantial and often unknown variability. Many more measurements than those presently available are, therefore, needed in order to enable oceanographers to understand the real nature of the oceans, their composition, stratification and motion.

14.2 Integrated Global Ocean Station System (IGOSS) and related activities

14.2.1 Ocean Data Stations

The value of a network of fixed oceanographic observing stations had been recognized by IOC since its foundation. This is not surprising with an international organization working in close contact with WMO, for which a global network of meteorological stations forms the observational basis of all operations. Although variability with regard to space and time is generally less pronounced in the ocean than in the atmosphere, recent investigations have clearly shown that oceanic processes are more variable than hitherto assumed and that the oceanic scales of motion are distinctly different from the atmospheric ones. Thus, it is evident that an oceanic network of fixed observing stations adequately spaced and occupied continuously would provide time-series of truly synoptic observations which would in turn form a unique basis for monitoring and modelling ocean processes with the aim of understanding the nature of the ocean.

IOC paid attention to this issue as early as its first session in 1961 when it asked its Bureau to establish a working group of experts from Member States, WMO and other appropriate international organizations which was to review the existing network of fixed oceanographic stations, study the needs for extending and improving it and prepare proposals for meeting these needs. The working group submitted a comprehensive report to the second session of the Commission in 1962, revealing the complexity and difficulty of the problem under study. First of all, the existing stations are of very different kinds. They include the following categories :

(1) Coastal and island stations ;
(2) Near-shore manned stations (occupied by light vessels, platforms, etc.) ;
(3) Offshore manned stations (occupied by ocean weather ships, etc.) ;
(4) Unmanned stations (occupied by automatic buoys, etc.) ;
(5) Stations on shipping routes taken by ships of opportunity ;
(6) Offshore reference stations visited regularly ;
(7) Cables in use for oceanographic observations ;
(8) Repetitive drifting observations (from ice islands, drifting buoys, etc.) ; and
(9) Satellite orbits.

As far as the location of these stations was concerned, a world chart was prepared showing the existing network of fixed oceanographic stations. This was published by IOC, together with a reference booklet, under the title of *Fixed Oceanographic Stations of the World, 1963.* This network was shown to be not homogeneous and very inadequate for many ocean areas.

The planning of a future network of ocean stations was considered entirely dependent on scientific and other purposes for which such a network would be needed. Should the data be collected generally, i.e. without a specific programme in mind, or for the solution of particular problems ? Most marine scientists favoured the second approach, but recognized that this would not be applicable when a permanent network was to be planned.

In addition to scientific problems connected with the objectives and desirable locations of such fixed oceanographic stations, a great variety of practical questions arose and were identified when further thought was devoted to the oceanographic network, including :

- Requirements regarding the volume and frequency of oceanographic observations made at fixed oceanographic stations ;
- Testing and intercomparison of instruments and methods in order to ensure comparable results ;
- Telecommunication and dissemination of those observations in real-time or near-real-time ;
- Reliable anchoring, distinctive marking, identification and legal status of unmanned automatic buoys and platforms occupying these stations ;
- Information for ocean-going shipping and fishermen about navigational hazards that such platforms might cause ;
- Protection of unmanned platforms against damage by shipping as well as against recovering or tampering by unauthorized people.

It was soon recognized that solutions of these manifold questions must be sought in co-operation with relevant international agencies, in particular with WMO, IMCO and Unesco. In the course of the further discussions the term «*ocean data stations*» was used from 1965 onward instead of «fixed oceanographic stations». The ocean data stations occupied presently are listed in the «International Catalogue of Ocean Data Stations» published as No 2 of the series *IOC Manuals and Guides* in 1975 with an Amendment No 1 in 1976.

14 .2.2 Communications

In order to deal with the important question of telecommunications in connexion with ocean data stations the Bureau was asked by the first session of IOC in 1961, to establish a *Working Group on Communications.* This group studied the problem in detail and, in 1962, submitted a report in which the allocation of specific radio channels was requested for the purpose of ocean data stations. As a minimum requirement, one 3.5 kHz channel in each of the six mobile maritime exclusive frequency bands in the 4, 6, 8, 12, 16 and 22 MHz ranges were considered necessary. In the following years these demands were discussed and further examined at meetings with adequate participation of national oceanographic and meteorological, as well as communication and maritime, authorities, in addition to representatives of the relevant international organizations. The efforts of IOC's Working Group on Communications were rewarded, when, in 1967,

Figure 12 A: Total number of observations exchanged (all types)

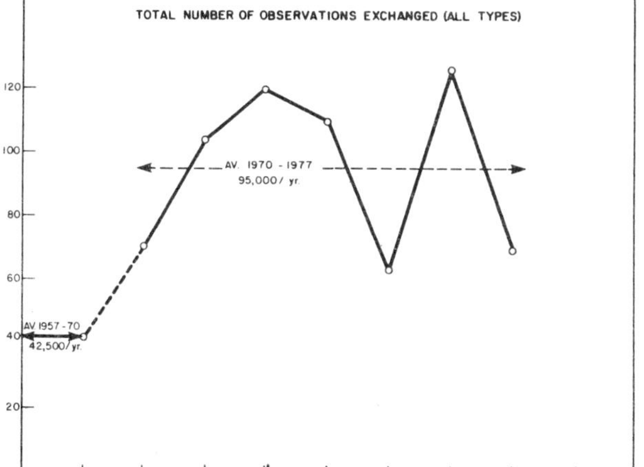

Figure 12 B: Number of oceanographic serial stations exchanged

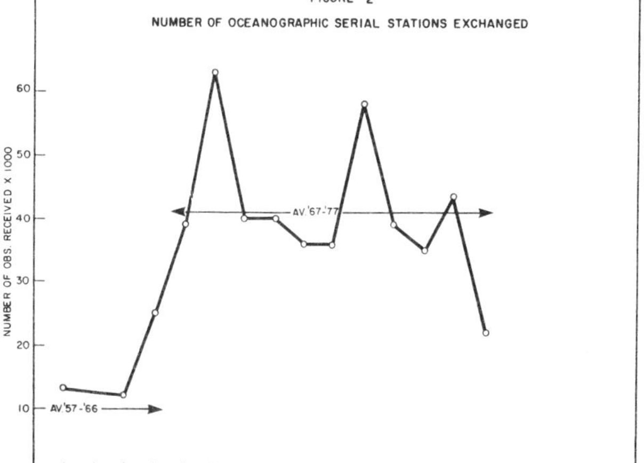

(from document IOC/EC-XI/14, Status of International Oceanographic Data Exchange, December 1978)

the World Administrative Radio Conference (WARC-67) designated the frequency bands as requested by the Commission for the transmission of oceanographic data. The WARC further requested IOC and WMO to prepare jointly, and in consultation with the International Frequency Registration Board (IFRB), a co-ordinated plan for the rational utilization of the allocated frequency bands in a system of ocean data stations established through international co-operation and to keep such a plan up-to-date in light of changing requirements.

For the successful pursuit of such work, a joint IOC/WMO Group of Experts on Telecommunications was thought to be more appropriate than the existing IOC working group consisting of intergovernmental representatives. Following a favourable reaction by WMO, this joint *IOC/WMO Group of Experts on Telecommunications* was established in 1968. It came to the conclusion that a well co-ordinated plan of using the frequencies must be based on requirements and on the development of detailed technical characteristics which were not yet fully known. The joint group, therefore, proposed a broad outline of the utilization plan and made preparations for the interim use of the frequencies allocated for ocean data transmissions, with the Secretariats of WMO and IOC assuming the co-ordinating function. This «*Agreed Interim Frequency Utilization Plan*» was first published in 1970 and has since been up-dated several times ; it has served its purpose well. The frequencies allocated have become increasingly used by ocean data buoys, as well as research vessels, ocean weather ships and certain voluntary observing ships for the transmission of oceanographic and meteorological messages.

In 1974, the World Maritime Administrative Radio Conference (WMARC) decided that the allocation of the six HF bands should be maintained without any change until the next Administrative Conference in 1979.

Regarding the exchange and dissemination of ocean data in real-time, it was decided that, to avoid the establishment of a duplicate system, the ocean material should be fed into the already existing Global Telecommunication System (GTS) of WMO via the National Meteorological Centres (NMCs) and exchanged in accordance with relevant WMO Procedures.

14.2.3 Legal status of Ocean Data Acquisition Systems (ODAS)

At its first session in 1961, the Commission recognized the need to clarify the legal status of manned and unmanned anchored buoys and platforms, i. e., the rights and duties of persons and states placing them in the sea. Having obtained a preliminary report on that subject, the second session of IOC in 1962 requested the Director-General of Unesco, in consultation with the Secretary-General of IMCO, to study the existing relevant international maritime conventions with a view to defining, in a new international convention, the legal status of such systems and the safety rules for their proper use. In partial response to this request, the Maritime Safety Committee of IMCO, together with the IOC Secretariat, drafted conclusions on the marking of data buoys which were approved by the IMCO Assembly. IOC, at its third session in 1964, recommended that all Member States should adopt and implement these conclusions.

The problem of the legal status of ODAS, however, remained unsolved. From the few national reports received by 1964 on that subject, IOC realized this to be a complex problem requiring careful study before any action could be taken to develop an international convention on such legal status.

In the following years, the Commission put much effort into clarifying the legal status of ODAS. In 1966, the IOC Bureau formed a group of experts in maritime law to assist the IOC Secretariat in preparing documentation dealing with national and international legislation and practices for a preparatory conference on the legal aspects of the use of ODAS. It was recommended that such a conference should be convened by Unesco and other international organizations, with the aim of preparing a draft international convention on the legal status of ODAS.

This group of experts met several times and, in close collaboration with IMCO, prepared a framework paper on the problems to be resolved when defining the legal status of ODAS, together with proposals for legal principles to be included in an international convention on the subject. This document was published as No 5 of the *IOC Technical Series* in 1969 under the title «Legal problems associated with Ocean Data Acquisition Systems (ODAS)».

The new term «*Ocean Data Acquisition Systems*» (ODAS) was introduced to cover the platforms, telemetering and non-telemetering buoys etc., used to collect data at ocean data stations of the categories (2), (3), (4) and (8) as defined in Section 14.2.1 above. The data collected can be of any type, although, at present, they are chiefly oceanographic and meteorological in nature.

The publication mentioned above provided a useful compilation of the principal legal problems arising from the use of Ocean Data Acquisition Systems – as analyzed in the light of the actual international law of the sea of that time. The following matters were considered :

- Rights of international organizations, governments and others to deploy Ocean Data Acquisition Systems (ODAS) ;
- Safety rules ;
- Liability of others in deliberate or accidental interference with ODAS ;
- Definition of the exclusive right of the owners of ODAS to the use of the devices and information obtained ;
- Legal responsibility for damage to others resulting from the use of ODAS ;
- Status of international organizations in the event of their developing programmes using ODAS.

The analysis given disclosed the uncertainty of the current international law concerning the operation of ODAS.

The group of experts further produced a preliminary draft convention on ODAS with technical annexes. All these documents were submitted to the joint Unesco/IMCO *Preparatory Conference of Governmental Experts to formulate a Draft Convention on the Legal Status of Ocean Data Acquisition Systems* (ODAS), which was eventually convened in 1972.

This Conference discussed most of the articles of the draft convention but did not succeed in reaching complete agreement in this respect. Nevertheless, the Conference made significant progress toward the preparation of such a draft convention in that it had been able to identify the main points of controversy. Thus, a second session of the Preparatory Conference was deemed to be necessary before the Plenipotentiary Conference of States could be called to adopt the ODAS Convention, which, probably will not be held until after the Third United Nations Conference on the Law of the Sea has completed its work.

Although the Preparatory Conference failed to agree upon a draft convention on the legal status of ODAS, it recommended advance implementation of certain safety provisions summarized in three technical annexes of this draft convention. Considering that the final convention on ODAS might not come into force for some time, the Maritime Safety Committee of IMCO and the Executive Council of IOC, acting on behalf of the Executive Board of Unesco, agreed to issue these three technical annexes, as amended by the Preparatory Conference, in a composite document, for Member States to use, on a voluntary basis, as guidelines for national measures. This document was published by Unesco and IMCO in 1972 under the title of *Safety Provisions of Ocean Data Acquisition Systems, Aids and Devices* (ODAS) and contains the following material :

I *Notification* — which provides for the procedure relating to the notification of deployment, activities and other information concerning ODAS.

II *Marking and Signals* — which prescribes provisions for day marking, lights and sound signals for different types of ODAS.

III *Construction Arrangements and Other Safety Provisions* — which specifies provisions for construction, fire protection, life-saving appliances, radio-communications and other safety aspects of ODAS.

14.2.4 Integrated Global Ocean Station System (IGOSS)

In the further planning of a network of ocean data stations, the aspect of user requirements received increased attention. Discussions were held with WMO with a view to forming a joint IOC/WMO Working Group on Ocean Data Stations to cope with this problem by uniting the interests and efforts of these two organizations. The whole subject was thoroughly investigated at the fifth session of the Commission in 1967 when the necessity was felt to streamline the structure and co-ordinate the work of IOC's subsidiary bodies related to an oceanographic network.

IOC realized that there was a growing requirement for oceanic data on a global scale by a wide spectrum of users comprising the fields of research, engineering, shipping and fisheries as well as the forecasting services for the general public. In response to this need the Commission decided to establish a permanent *Working Committee for the Integrated Global Ocean Station System (IGOSS)*. This Committee was given the task of planning and co-ordinating a relevant IOC programme of studies and services in oceanic areas both within the Commission and jointly with WMO.

The existing working groups on communications, variability, and ocean-atmosphere interaction were re-structured into a co-ordinated system subsidiary to the new Working Committee. This new organization was considered as a big step forward in the development of ocean services. The creation of IGOSS was certainly influenced by WMO's efforts to strengthen its global observational system of World Weather Watch (WWW) wherein oceanic observations form a highly important part. On the other hand, IGOSS constitutes a substantial component of the Long-term and Expanded Programme of Oceanic exploration and Research (LEPOR), formulated by IOC on request by the United Nations and subsequently approved by its General Assembly in 1969.

Following this action of IOC, WMO established an Executive Committee *Panel on Meteorological Aspects of Ocean Affairs* (MAOA) with the task of effecting the closest collaboration with the IOC Working Committee for IGOSS and to ensure consistency in the planning and implementation of IGOSS on the one hand and of WWW on the other hand. The meetings of the two bodies WC/IGOSS and EC/MAOA were held jointly in most cases.

At its sixth session in 1969, IOC adopted the «*General Plan and Implementation Programme for IGOSS for Phase I*». This approach, proceeding in several consecutive phases, was taken in view of the magnitude of the ultimate goals of IGOSS and of the urgent need for an initial plan of action. Phase I of the implementation programme for IGOSS was based on existing technology and carefully correlated with the WWW programme of WMO. Phase II was to incorporate new technology, e.g. satellites and buoys. When developing this programme, IOC was well aware of the need for more detailed studies on :

- the oceanographic requirements regarding the location of observing stations and of the parameters to be observed, including their accuracy ;
- the requirements of users on oceanographic information.

The internal structure of the Working Committee for IGOSS was re-arranged by IOC at its seventh session in 1971 and subsequently consisted of the following bodies :

- *Joint IOC/WMO Planning Group for IGOSS (IPLAN)*, functioning as a steering committee for IGOSS ;
- *Joint IOC/WMO Group of Experts on IGOSS Technical Systems Design and Development and Service Requirements (ITECH)* ;
- *Joint WMO/IOC Group of Experts on Telecommunications (ITEL)* ;
- *Group of Experts on Oceanographic Research as it relates to IGOSS (IRES)*.

Later, when rationalizing the stucture of its subsidiary bodies, the IOC, at the eighth session of its Assembly in 1973, re-organized IGOSS again by combining all groups of experts mentioned above into a *Joint IOC/WMO Group of Experts* so that specific groupings could be formed to deal with special scientific and technical matters. The IOC/WMO Planning Group for IGOSS (IPLAN) was retained.

In 1977, the Commission, at the tenth session of its Assembly and in concurrence with WMO, decided to strengthen the co-operation with WMO even more and, to this effect, to replace both the IOC Working Committee for IGOSS and the Joint IOC/WMO Planning Group for IGOSS by a *Joint IOC/WMO Working Committee for IGOSS*, which is open to any member of IOC and WMO wishing to participate in any part of the IGOSS programme.

14.2.5 Development of the IGOSS Programme

In the initial phase, the Working Committee for IGOSS concentrated on developing the necessary operational tools and working procedures. New codes for the transmission of oceanographic data were adopted, i.e.

- the BATHY code for temperature and depth values (as obtained by bathythermographs) ; and

– the TESAC code for values of temperature, salinity, current and depth.

Further, operational instructions were issued in the following manuals and guides :

- «Manual on IGOSS Data Archiving and Exchange» (*IOC Manuals and Guides* No 1) ;
- «Guide to Operational Procedures for Collection and Exchange of Oceanographic Data (BATHY and TESAC)» (*IOC Manuals and Guides* No 3) ;
- «Guide to Oceanographic and Marine Meteorological Instruments and Observing Practices» (*IOC Manuals and Guides* No 4).

The first operational programme of IGOSS was the *BATHY Pilot Project*, launched in 1972. This programme was designed for the collection, exchange and evaluation of bathythermograph data and served to test and evaluate the related systems. The number of BATHY/TESAC reports exchanged in real-time or near-real-time via the Global Telecommunication System (GTS) amounted to 18,600 in 1973 and to 15,000 in 1974. These totals were already a significant portion (25%) of the total number (58,083 temperature and salinity observations from oceanographic casts) received by World Data Centre 'A' (Oceanography) in 1973. Thus the BATHY pilot project showed the great potential for increasing the amount of temperature data to be exchanged on a synoptic basis if more countries participated in this project. In anticipation of increased participation, the BATHY Pilot Project was converted into an operational programme in 1975. Such confidence was rewarded in that the number of BATHY/TESAC reports exchanged over GTS increased to 16,500 in 1975, to 33,500 in 1976 and to 38,200 in 1977. The development of the BATHY/TESAC exchange within IGOSS through the years from 1972 to 1977 is illustrated in Fig. 13.

Recognizing that developments in ocean services have necessitated some changes in the scope of IGOSS, the Working Committee for IGOSS, together with the WMO/EC Panel of Experts on Meteorological Aspects of Ocean Affairs (MAOA), prepared a new «General Plan and Implementation Programme for IGOSS». This plan, which was approved by the IOC Executive Council and the WMO Executive Committee in 1976, is concerned with the period from 1977 to 1982. This second General Plan, which incorporates new techniques and results of experience, is intended to serve as a guide for the further development of IGOSS and is governed again by the central principal that the WMO Marine Meteorological Services and IGOSS are complementary and should, thus, be developed and operated together.

The building-up of IGOSS has now progressed to a stage where the basic elements of the system can be identified as follows :

- The *IGOSS Observing System (IOS)*, consisting of oceanographic and marine meteorological observations from ships, buoys, satellites and other platforms ;
- The *IGOSS Telecommunication Arrangements*, based on the use of the WWW Global Telecommunication System (GTS), satellite relay or interrogation links or other new techniques for collection and distribution of observational data from oceanic platforms as well as for exchange and distribution of processed data ;
- The *IGOSS Data Processing and Services System (IDPSS)*, consisting of national, specialized and world oceanographic centres for the processing of observational data and preparation of oceanographic products (analyses and forecasts) and provision of services to various groups of users. A review of «Oceanographic products and methods of analysis and prediction» was published in 1977 as No 12 of the *IOC Technical Series*. It comprises information of such products as sea surface and subsurface temperature, sea state, water level, storm surge, surface current and other parameters mostly disseminated in the form of charts.
- The *IGOSS Data Archiving and Exchange System*, using existing mechanisms and channels for international oceanographic data exchange as developed by the IOC Working Committee on IODE.

In addition to the basic elements mentioned above, IGOSS has embarked on the IGOSS Marine Pollution Monitoring Programme in response to recommendations of the Stockholm Conference on the Human Environment, supported by IOC and WMO. This function of IGOSS is co-ordinated with the IOC Working Committee for GIPME (cf. Section 11.3).

Further to these prominent operational elements of IGOSS, there exist further programme components which either belong :

- to the activities of training, education and mutual assistance (TEMA), in order to enable developing countries to participate actively in the IGOSS programme ; or
- to the area of research which is needed in order to ensure a sound scientific basis for the development of IGOSS.

The main motivation for carrying out ocean services is the potential benefit which can be expected from them. In the case of IGOSS, such benefit covers a wide range, since the need for actual information on the ocean environment goes beyond the oceanographic community and encompasses scientific and industrial institutions as well as governments, and public and private interests concerned with marine affairs. Table 8 summarizes the users, the kind of oceanographic information needed by them and also the potential benefit of such IGOSS products.

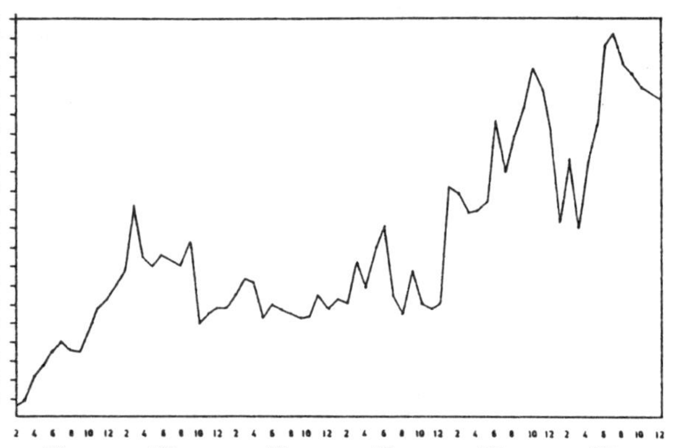

Figure 13 - Total number of BATHY/TESAC data exchanged monthly within IGOSS from 1972 to 1977 (from IOC Biennial Report for 1976-1977).

Table 8 Users, contents and potential benefit of IGOSS products

User groups	Contents of oceanographic information	Areas which may benefit by oceanographic information
Coastal population	storm surges, tsunamis	early warning, protection and safety
Sea fisheries Coastal fisheries and mariculture	temperature, salinity currents, sea state tides and tidal streams, storm surges	improved efficiency, exploitation and management, protection and safety
Shipping	storm surges, tsunamis tides, tidal streams currents, sea state, temperature	protection and safety improved economical operation navigation, routeing, cargo care
Ocean, offshore and coastal engineering including industrial operations	temperature, salinity, currents, tides and tidal streams, waves, storm surges, tsunamis	improved efficiency, design, operational planning and management, exploitation of mineral resources, protection and safety
Pollution abatement and control	temperature, salinity, currents, tides, tidal streams	protection of living resources and man, effluent and waste disposal ; distribution, transport, decomposition of pollutants
Sea ice and iceberg warning and prediction	temperature, salinity, currents	formation and break-up, movement and decay
Search and rescue operations	temperature, currents, tides, tidal streams, storm surges, tsunamis	improved efficiency
Harbour construction and management	tides, tidal streams, tsunamis, range action	design, scheduling, management, protection and safety
Marine recreation, yachting	temperature, currents tides, tidal streams, storm surges, tsunamis	planning, navigation protection and safety
Weather forecasting Climate modelling	surface temperature distribution ; stratification and dynamics of the ocean	improved short-term forecasts of weather and anomalous radio propagation ; coupled ocean-atmosphere models
Oceanographic research programmes	real-time information on stratification, composition and motion of water masses in research area	assistance for decision-making during field operations
Bathymetric services, hydrographic surveys	tides, tidal streams	improved accuracy

14.2.6 Future of IGOSS

The area coverage achieved by the IGOSS programme in 1977 is shown by Fig. 14 summarizing bathythermograph data over a period of six months.

It is well understood that to implement IGOSS on a fully global basis will take many years of effort. As a comparison, the operational global meteorological networks have been under development for the last 100 years and still are not fully implemented in some land areas. However substantial progress has been made in developing ocean services programmes in many nations, and these growing national programmes are continuing to strengthen the global objectives of IGOSS.

During the last few years, particular emphasis has been placed on developing projects in support of practical marine activities such as fisheries (e.g. ICNAF Flemish Cap Experiment) and of scientific studies and experiments (e.g. ICES OVERFLOW-73, FGGE, POLYMODE, climate programme). Studies have been initiated to enhance IGOSS through the concentrated efforts of Member States in certain oceanic areas, where practical and scientific tasks can be more clearly identified, co-operative programmes can be established, and where sufficient data coverage can be expected. Particular attention has also been given to further study of requirements of various marine users for oceanographic services and exchange of knowledge and experience among specialists on oceanographic product formulation and methods used for their preparation. This emphasis on IGOSS products and IGOSS user requirements brought about the preparation of 17 IGOSS products in response to user needs (both scientific and operational) for the First GARP Global Experiment (FGGE). It is anticipated that

Figure 14 - IGOSS bathythermograph locate plot, January-June 1977
Total number of observations: 8,027 (from IOC Biennial Report for 1976-1977).

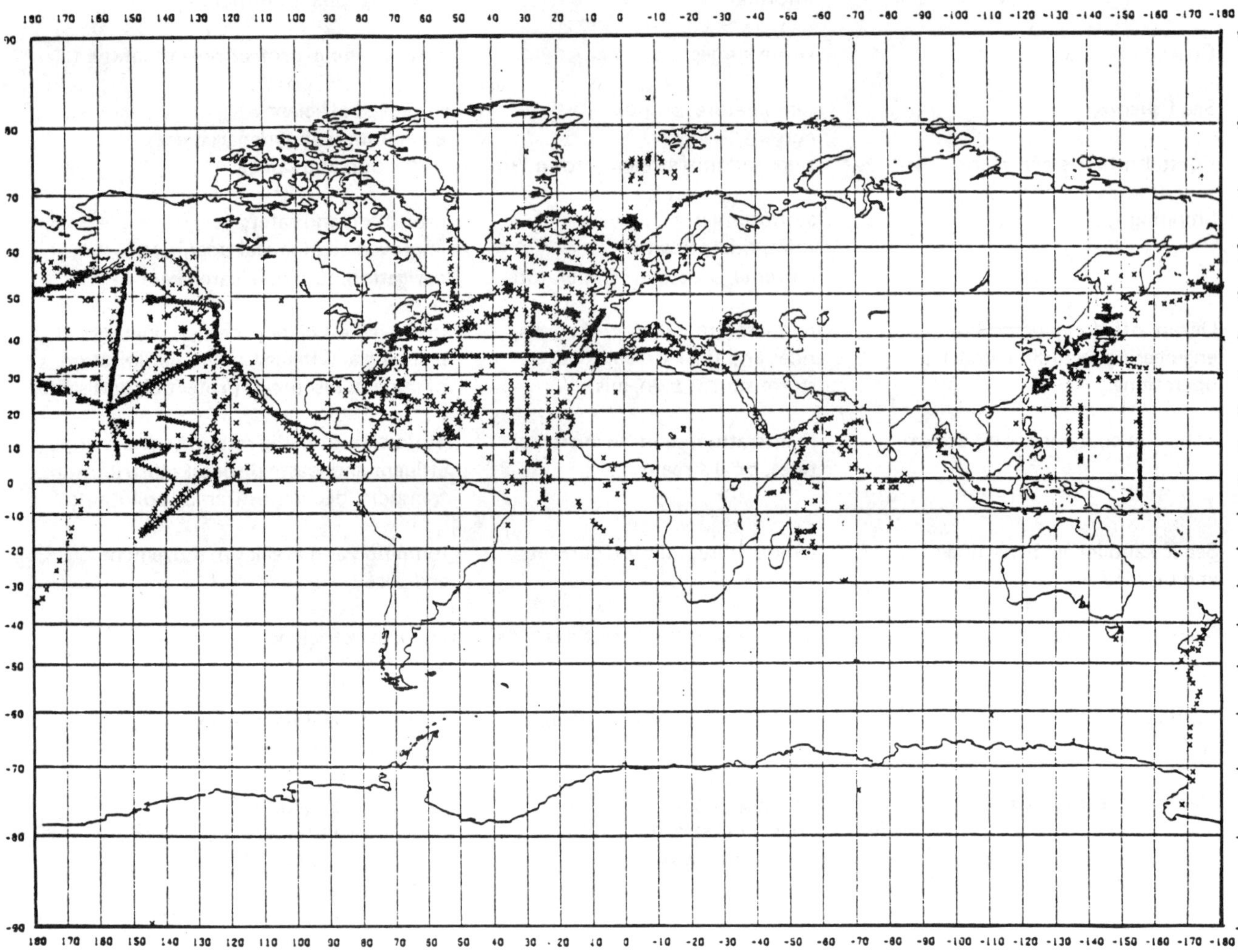

these IGOSS products for FGGE will not only lead to a considerable increase during FGGE, but will also further establish the viability of the IGOSS programme on a permanent basis.

Future work will also concentrate on increasing the input from voluntary observing ships and the incorporation into the IGOSS Observing System of new data acquisition devices such as buoys, satellites and radars with automatic sensing and data transmission. In addition, the basic understanding of the physical processes going on in the ocean will also be explored with a view to developing predictive models which can be used for forecasting oceanic conditions using IGOSS data.

Considering the vast ocean areas, an efficient global oceanographic service programme encompassing the various components described above is a substantial undertaking that can only be tackled and carried out successfully if there exists substantial interest on a multinational level.

With the growing interest in ocean monitoring systems for climate, fisheries operation, shipping, research, etc., the IGOSS programme is attracting this multi-national interest, and will continue to seek further co-operation among all IOC Member States, including the developing countries, as well as close co-operation with WMO and other international organizations.

14.3 Marine pollution monitoring

In 1970, when rendering advice to IOC upon the scope and content of LEPOR, the Group of Experts on Long-term Scientific Policy and Planning (GELTSPAP) pointed to the necessity of «identifying the elements of a desirable and feasible marine pollution monitoring system». The term «monitoring» was defined by the Preparatory Committee for the UN Conference on the Human Environment as «a system of continued observation, measurement and evaluation for defined purposes». In the case of marine pollution these purposes are aimed at :

- providing early warning of significant environmental changes to induce consideration of protective measures ; and
- enhancing quantitative knowledge of the marine environment, including the means by which dynamic balance is maintained in marine ecosystems, as a basis for managing marine resources.

Further, it is necessary to determine which pollutants and marine materials are to be monitored and how. In this respect, GELTSPAP recommended that such a monitoring programme should be focussed on the physical, chemical and biological concentration mechanisms that facilitate detection as well as determine the significance of marine pollutants.

IOC, at its seventh session in 1971, reviewed this problem and concluded that a marine pollution monitoring programme should be developed as a function of IGOSS as far as physical and some chemical properties are concerned. The Commission further recognized that marine pollution monitoring includes substantial amounts of research because methods for monitoring many chemical properties and biological processes had at that time yet to be developed. Such work would fall under the responsibility of the GIPME programme (see Section 11.3). Thus, two subsidiary bodies of IOC, the Working Committee for IGOSS and the International Co-ordination Group (later Working Committee) for GIPME, were mainly engaged in planning and implementing the marine pollution monitoring programme of IOC.

Such approach was completely in line with resolution 90 of the UN Conference on the Human Environment, endorsed by the UN General Assembly, which recommended that «IOC, jointly with WMO and in co-operation with other interested intergovernmental bodies, should organize marine pollution monitoring, preferably within the framework of IGOSS».

In the light of the above decisions, the Working Committee for IGOSS, in particular its Joint IOC/WMO Planning Group, went to work and recommended that, as a first step in the development of a Marine Pollution Monitoring Programme, a Pilot Project should be undertaken and addressed to parameters that are fairly easily monitored with the widely available present technology. In this context, the group focussed its attention on the monitoring of oil, dissolved petroleum constituents and particulate petroleum residues. A relevant operational plan for a *Marine Pollution (Petroleum) Monitoring Pilot Project (MAPMOPP)* was developed and submitted for approval to the eighth session of the IOC Assembly in 1973. The Commission endorsed this Pilot Project, as did the WMO Executive Committee. UNEP was asked to support it and responded favourably. The main objective of the Pilot Project was to determine the feasibility of developing and operating such a programme on a global scale, with the ultimate goal being to obtain information on the distribution and dynamics of petroleum in the world ocean.

The monitoring activities included :

- visual observations of oil slicks, petroleum residues and other floating pollutants (e.g. plastic materials) ;
- collecting surface samples for later inspection in laboratories with respect to particulate petroleum residues (tar balls) ;
- tar sampling on beaches ; and
- taking water samples for later analysis in laboratories with regard to petroleum hydrocarbons dissolved and dispersed in sea water.

The observations or samples taken were to be accompanied by complementary information on environmental conditions.

The areas for the Pilot Project originally chosen, took into account existing national and regional programmes, regions of offshore oil production, the main routes of oil transportation, and main ocean currents. However, during the first years of the development of the project, the areas of monitoring were finally extended to include any ocean area in order to engage as many countries as possible and in order to base it on any regional activities being set up. This coverage of the world ocean is understood as a preparatory step for possible later phases of global pollution monitoring activities.

Visual observations were carried out from a great variety of ships, fishing vessels, offshore platforms and aircraft, while the taking of samples was left to special ships such as ocean weather ships and research vessels, and coastal stations.

An international Symposium and Workshop on Marine Pollution (Petroleum) Monitoring was held in Gaithersburg, USA, in 1974 to provide scientific and technical advice as to the methodology to be used during the Pilot Project, as well as to recommend technical assistance and training programmes to developing countries. Scientific guidance on the implementation and further development of the Marine Pollution Monitoring Programme within IGOSS was furnished by the Working Committee for GIPME, whereas the Working Committees on IODE and for TEMA rendered advice with regard to data management and needs for training, education and mutual assistance.

The implementation of the operational phase of the Pilot Project started on 1 January 1975 and was scheduled to last for two years. A substantial number of Member States indicated their interest and their willingness to participate in the project. Thirty six countries nominated National Co-ordinators for the Pilot Project and 12 Member States sent reports on their relevant activities.

In order to review the progress reached, a Second IOC/WMO Workshop on Marine Pollution (Petroleum) Monitoring was convened in 1976 in Monaco. The evaluation showed that the Pilot Project was capable of providing large amounts of data, in particular visual observations of oil slicks. Fewer data were, however, available on dissolved, dispersed, floating and stranded oil residues.

In general, the validity of the original objective and the usefulness of the Pilot Project were confirmed, although there was felt to be an urgent need for a thorough scientific evaluation as to how far the data gathered within the project do in fact reveal the actual extent of petroleum pollution in the ocean. Further, broader participation, including countries with limited capabilities, was encouraged. Relevant arrangements for training and technical assistance were given high priority. The workshop recommended that the Pilot Project be extended until the end of 1978.

A revised version of the operational plan was published under the title of «Guide on Operational Procedures for the IGOSS Pilot Project on Marine Pollution (Petroleum) Monitoring» (*IOC Manuals and Guides*, no 7 (1976)).

An IGOSS Sub-group of Experts has been responsible for periodic review of the progress of the Pilot Project and for evaluating its results. At its second session, the Sub-group concluded that in an eventual future monitoring programme, the elements «visual observations of oil» and «floating tar balls» could be included, but that the other two parameters of the project should be subject to further evaluation before a decision be made on their inclusion.

The Sub-group decided to review the project in mid-1979 and that a final decision on its conversion into an operational monitoring programme would be made at a Third Workshop on Marine Pollution (Petroleum) Monitoring to be convened at the end of 1979.

In addition to the monitoring of pollution caused by petroleum, the broader aspects of marine pollution as originating from other deleterious substances were also considered. Apart from a number of marine pollution monitoring programmes carried out at the national or regional levels, IOC, in co-operation with WMO and UNEP, approved a pilot project for monitoring background levels of selected pollutants in open-ocean waters in order to establish a mechanism for measuring secular trends in the pollution of the world ocean, particularly in areas far away from the sources of pollutants.

14.4 Tsunami Warning System in the Pacific (ITSU)

When, in the Pacific Ocean, an earthquake occurs below or near the ocean floor, a series of extremely long ocean waves may be created which have become known by the name of «tsunami». The length from crest to crest of such waves, travelling at speeds exceeding 1,000 kilometres per hour, may reach several hundred kilometres, whereas its height in deep water is only about one metre. Consequently, the waves cannot be recognized either from a ship or from an aircraft. But when they approach a coast and enter shoaling water, their length and speed decrease with a simultaneous large increase of the height of their crests, which may reach heights of more than 35 metres. Such huge waves have disastrous effect on beaches, harbours and coastal facilities. Therefore, an early warning is of great importance for the protection of the life and property of the coastal population.

Such a tsunami warning system was established by the United States in 1948. It consists of a number of seismological and tidal stations around the Pacific Ocean and operates a headquarters in Honolulu, Hawaii. Once an earthquake of sufficient strength to generate a tsunami is observed in the Pacific Ocean area, the location of its epicentre is determined and, if this site is such that a tsunami may develop, a tsunami watch is issued which alerts all participants of the system. If the generation of a tsunami is confirmed by tidal records taken in the neighbourhood of the epicentre, a tsunami warning follows with predicted arrival times for given locations in the Pacific Ocean. Since the propagation of these long ocean waves adheres to known physical laws, accurate arrival times can be estimated for these potentially destructive sea waves.

The subject of the Tsunami Warning System was first brought to the attention of IOC at its second session in 1962. At that time, 10 countries co-operated with the USA in the operation of the Tsunami Warning System, thus making it an international undertaking. But further improvement, e.g. by establishing additional observation posts for obtaining seismic and sea level data, was both needed and feasible. IOC, therefore, recommended that Member States in the Pacific area, which did not yet take part in the present system, establish adequate internal structures regarding communications, seismic and tidal stations and integrate their system with that operated by the United States. In addition, increased research on the causes, nature and effects of tsunami was encouraged.

At its fourth session in 1965, IOC took more formal action, based on the comprehensive report of a working group on that matter. It recognized the existing national Tsunami Warning Centre in Honolulu as the *International Tsunami Information Center (ITIC)* of the Commission and established an International Co-ordination Group composed of interested Member States in the Pacific area as the responsible subsidiary body of IOC to take care of all aspects of the Tsunami Warning System, in particular to effect liaison among the participating Member States and with other interested organizations.

Membership of the *International Co-ordination Group for the Tsunami Warning System in the Pacific* has increased from 6 in 1968, to 14 in 1971, to 18 as of December 1978. Correspondingly, the system was substantially expanded and improved. Much attention was given to aspects of education, training and mutual assistance in order to help developing countries with long coastlines or subject to danger from tsunamis in the building-up of the necessary infrastructure. Further, educational material was provided in order to create public awareness of the potential dangers and protective actions that should be taken upon receipt of a tsunami warning. The whole Tsunami Warning System, as it was in 1973, is illustrated in Fig. 15.

In the course of time, the activities of the International Tsunami Information Center (ITIC) of IOC broadened in scope so that its mandate and functions had to be reformulated. It works closely together with the US National Tsunami Warning Center in Honolulu, and its mission is to mitigate the effect of tsunamis throughout the Pacific Ocean by :

- monitoring the international tsunami warning activities and recommending improvements regarding their performance ;
- providing assistance in the establishment of national warning systems ;
- promoting tsunami research and its application so as to prevent loss of life and damage to property ; and
- co-operating with the World Data Centres in making available all records pertaining to tsunamis.

14.5 General Bathymetric Chart of the Oceans (GEBCO)

Bathymetric charts of the ocean representing the morphology of the sea floor at appropriate scales are an indispensible element in the management of ocean affairs. This is especially true for scientific investigations, as all marine disciplines have a strong interest in reconnaissance and detailed charts of the ocean bottom, as a basis for further research. Such charts are also of very great practical value in that they are required for all aspects of mineral exploitation, fisheries, engineering construction, and other operations on or above the sea floor. The importance of precise bathymetric charts will increase even more when practical consequences have to be drawn from the regulations of a future convention on the law of the sea.

Guided by such arguments, IOC, at its seventh session in 1971, included morphological charting of the sea floor as one of the eight programmes of major importance in its Long-term and Expanded Programme of Oceanic exploration and Research (LEPOR). Such a programme, although containing certain features of research, should be considered a service, because its chief purpose is to serve the scientific community and the public by providing the best information available on the configuration of the ocean bottom.

This decision must be seen in the context of the project known as the «*General Bathymetric Chart of the Oceans*» (GEBCO) which was established in 1903 by Prince Albert I of Monaco and had been the responsibility of the International Hydrographic Bureau (IHB) in Monaco since its foundation in 1921. This chart series consisted of 24

Figure 15

Tsunami warning system in the Pacific Ocean

Participating Member States

Canada
Chile
China
Ecuador
France
Guatemala
Indonesia
Japan
Korea (Rep. of)
New Zealand
Peru
Philippines
Singapore
Thailand
USSR
USA

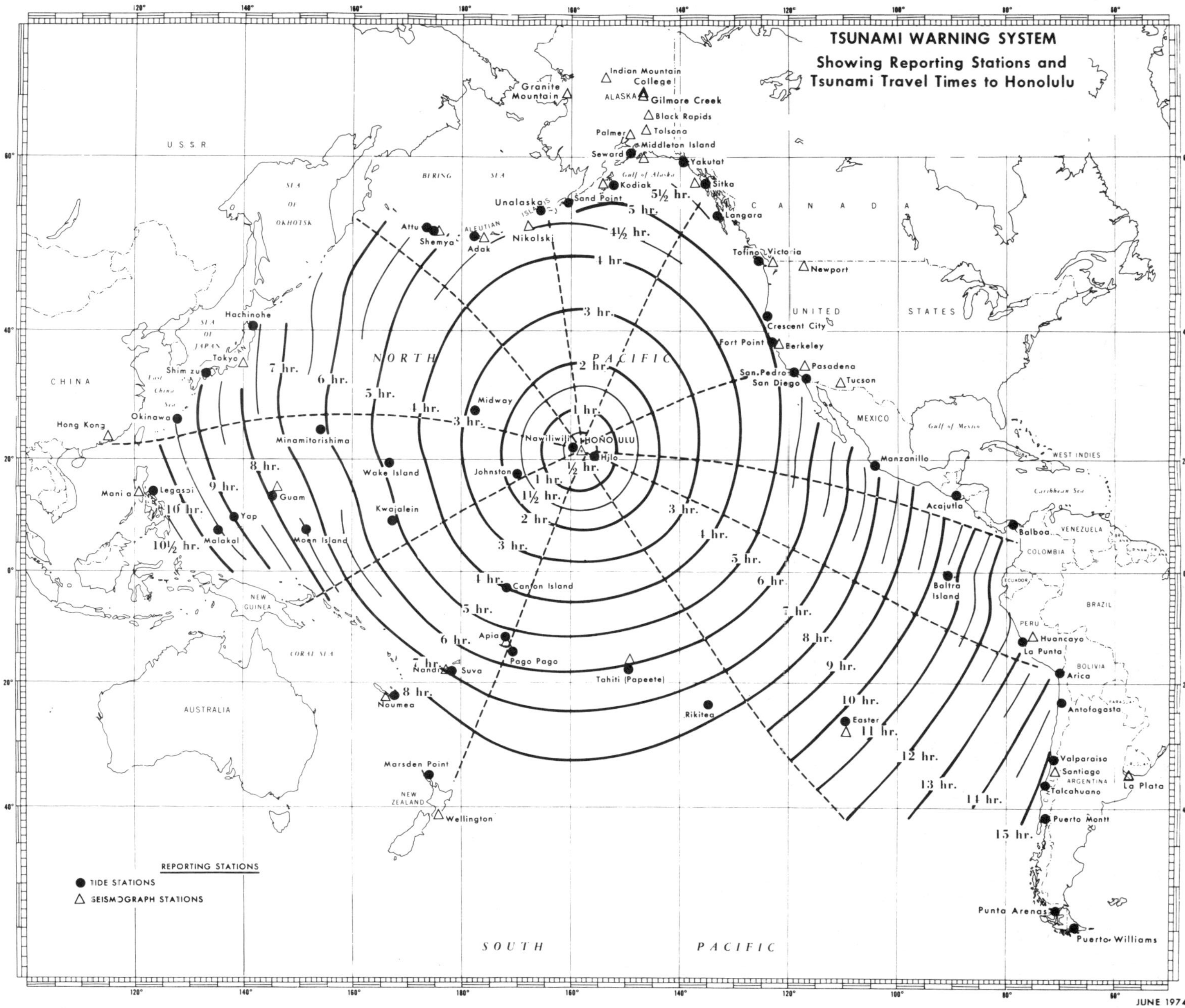

sheets covering the whole globe (16 sheets on Mercator projection on a scale of 1 : 10 million at the equator and 8 polar sheets). The bathymetric data were collected and assembled on master plotting sheets on the scale of 1 : 1 million by national hydrographic offices volunteering to take responsibility as regional centres for certain oceanic areas. The GEBCO sheets were then compiled from these plotting sheets and published by the International Hydrographic Bureau with necessary scientific advice being provided by a committee of representatives from ICSU, SCOR, IHB and the International Association of Physical Oceanography (IAPO), (later renamed the International Association for the Physical Sciences of the Ocean (IAPSO)).

In the course of time, this well-established scheme ran into financial difficulties. The sheets could not be up-dated regularly and the rate of production decreased.

IOC, at the third meeting of its Bureau with the Consultative Council in 1963, reviewed the situation and stressed the great scientific importance of expeditious preparation of bathymetric charts of the world ocean based on available soundings. The Bureau urged the Member States of IOC to support vigorously the efforts of their hydrographic services in the task of maintaining the plotting sheets. A similar resolution was adopted by the Commission at its third session.

In 1964, the situation was improved in that the IHB accepted an offer from the hydrographic office of France (Service hydrographique de la Marine) to assume the financial and technical responsibility for the final publication of the 1 : 10 million GEBCO sheets, and the fourth edition of GEBCO was started in co-operation with the Institut géographique national de France. IOC's contribution to this new edition consisted of asking its Member States to assist the IHB in eliminating doubtful bathymetric data from navigational charts. Oceanographic cruises should investigate unconfirmed soundings whenever possible. A comprehensive list of such requirements was prepared by the IHB on request by the IOC.

In the years 1972 and 1973, the SCOR Working Group on Morphological Mapping of the Ocean Floor critically reviewed the GEBCO programme, particularly with respect to its scientific value. The group emphasized that the contouring of sounding data should be done with due regard to the existing state of knowledge of sea floor morphology and of the geological and geophysical processes active at the ocean bottom. Further, the state of production at that time of the 24 GEBCO sheets was considered unsatisfactory, it being noted that the publication dates of the different sheets ranged from 1923 to 1970 and only 8 out 24 had been published after 1960. This considerably decreased its value to scientists, and this was reflected in the low sales. Following a long discussion, ideal specifications for a new world bathymetric chart series were prepared which formed the basis for the subsequent action of IOC and the International Hydrographic Organization (IHO) (formerly IHB).

In response to the proposals of SCOR, the Commission, at the eighth session of its Assembly in 1973, approved the establishment of a *Joint IOC/IHO Guiding Committee for GEBCO* after consultation with SCOR, the International Association for the Physical Sciences of the Ocean (IAPSO) and the Commission for Marine Geology (CMG) of the International Union of Geological Sciences (IUGS), in order to replace the former GEBCO committee. This new guiding committee was given the task of :

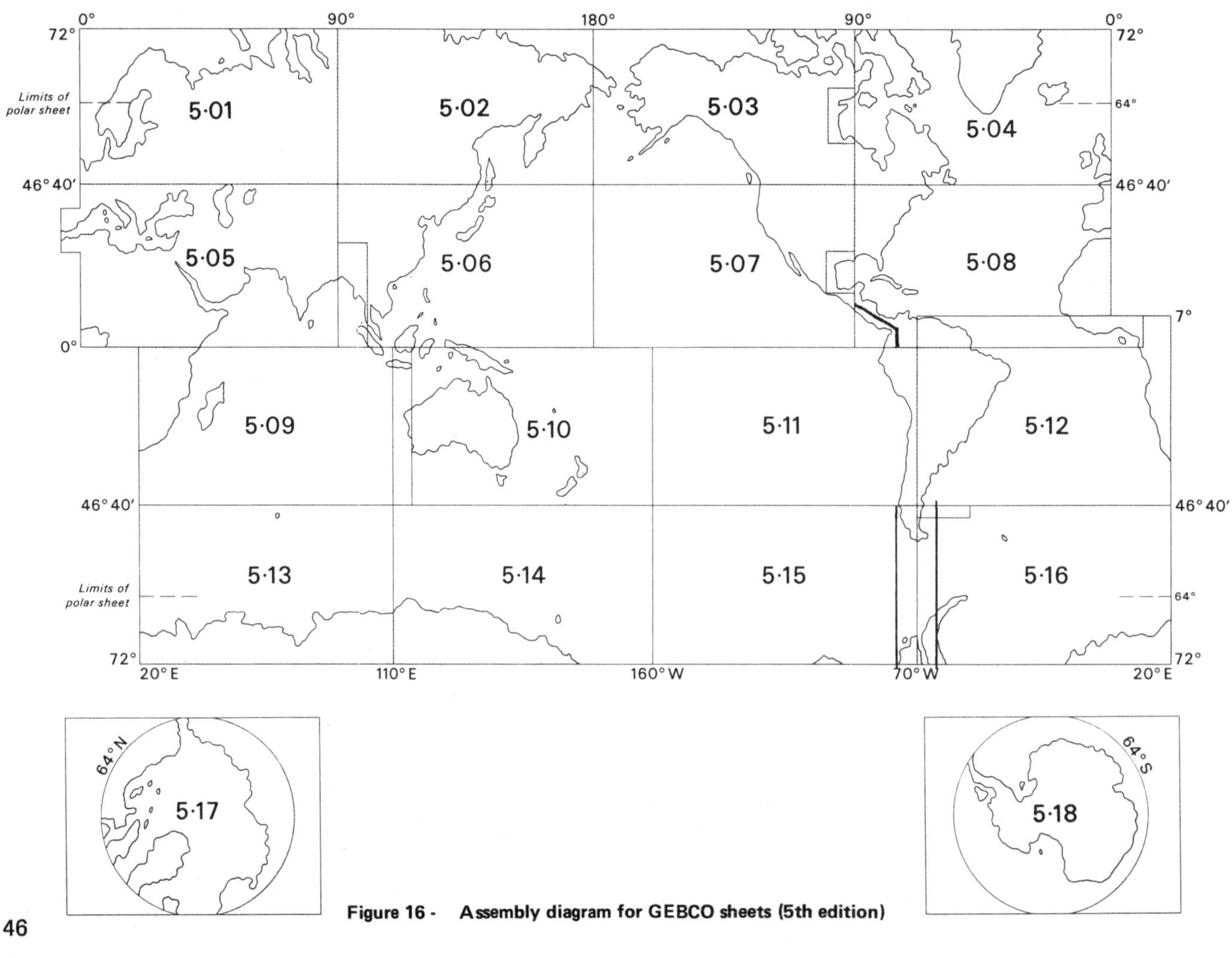

Figure 16 - Assembly diagram for GEBCO sheets (5th edition)

- determining the needs of the scientific community, educational authorities and other users for GEBCO charts ; and of
- producing, based on these needs, new specifications for the preparation and production of the world bathymetric 1 : 10 million series charts, within the general guidelines developed by the SCOR Working Group.

The role of IHO as the specialized World Data Centre for Bathymetry was to be maintained.

The Joint IOC/IHO Guiding Committee tackled its work immediately and developed the plan for the fifth edition of GEBCO with the aim to place speedily at the disposal of the user community the best available interpretation of bathymetric data of the world ocean. By rearrangement of the sheets of the polar regions the number of 1 : 10 million sheets was reduced from 24 to 18 (see Fig. 16). The Guiding Committee further proposed that an International GEBCO Geoscience Unit be established with the principal responsibility to prepare contour charts of the GEBCO 1 : 10 million series, under the general guidance of scientific co-ordinators nominated by the GEBCO Guiding Committee.

The first sheet of the new GEBCO edition was published in 1975 and the hope was expressed that world coverage would be achieved within a few years. The Canadian Hydrographic Service agreed to publish the first four sheets. To date, funds have still not been found to support the proposed Geoscience Unit. Although a review of the progress made on each of the individual GEBCO sheets showed that the compilation and production schedule was in a thoroughly healthy state, it appeared unlikely that the fifth edition of GEBCO would be completely published before 1982.

15 Scientific information and evaluation

In order to fulfil its task as a promoting and co-ordinating body in the field of marine science, IOC must keep abreast of new developments in ocean research. One way of achieving this is to provide for lectures on suitable topics, given by eminent marine scientists, at sessions of the IOC Assembly and, possibly, at those of its Executive Council.

There are different possibilities of selecting such themes, namely :

- Presenting integrated scientific reports on the results of co-operative investigations organized and co-ordinated by IOC, i. e., a retrospective review ;
- Elucidating scientific programme proposed for co-operative investigations ;
- Taking a forward look by discussing topics of importance for future programming of co-operative investigations of the Commission ; and
- Delivering a lecture in honour of a famous oceanographer.

IOC has made arrangements for lectures of all four categories. At the first session of the Commission in 1961, Prof. A.D. Dobrovolsky (USSR) gave a lecture honouring the memory of the great explorer, oceanographer and humanist Fritjof Nansen.

After the Commission had lost its first Chairman, Professor Anton Bruun, who died on 13 December 1961 during his term of office, the IOC Bureau decided that, at each session of the IOC, an evening lecture be dedicated to the memory of the first Chairman of the Commission. As a consequence of this decision, the following *Bruun Memorial Lectures* were given :

IOC-II	1962 :	The question of convection in the earth' s mantle by Dr. Roger Revelle (USA)
IOC-III	1964 :	New light on the deep-sea fauna by Dr. Torben Wolff (Denmark)
IOC-IV	1965 :	Variability in the ocean by Prof. R. W. Stewart (Canada)
IOC-V	1967 :	Air-sea interaction : a forward look by Prof. H. Charnock (UK)
IOC-VI	1969 :	Sea-floor spreading in the South Atlantic (Expedition of the Glomar Challenger) by Dr. A.E. Maxwell (USA)

At its sixth session in 1969, the Commission reviewed this practice and decided that in view of the growing complexity of its scientific engagement, at future sessions, three Bruun Memorial Lectures should be held, summarizing important developments during the preeceding two-year period in the following fields :

- Geology, geochemistry and geophysics ;
- Physical and chemical oceanography as well as meteorology ;
- Marine biology.

Furthermore, arrangements for regular publication of these lectures were deemed necessary.

Following this decision, lectures on the scientific results of the International Indian Ocean Expedition (IIOE) were given at the seventh session of IOC in 1971, namely :

- Meteorological results of the IIOE by Dr. C.S. Ramage (USA),
- Physical results of the IIOE by Dr. J.C. Swallow (UK),
- Biological results of the IIOE by Dr. G.F. Humphrey (Australia),
- Geological results of the IIOE by Dr. G.B. Udintsev (USSR) (announced but not presented).

These lectures were published in *IOC Technical Series* No 10 (1972). At the same session the IOC expressed the desire to be kept even more informed about the problems and progress of marine science throughout the world and authorized the Executive Council to take care of suitable arrangements at sessions of the Commission and/or its bodies.

As a consequence of this decision, IOC experienced a wealth of scientific information at the eighth session of its Assembly in 1973. First, some information was offered on the forthcoming Global Atmospheric Research Programme (GARP) of WMO and ICSU as well as on its sub-projects – the GARP Atlantic Tropical Experiment (GATE) and the First GARP Global Experiment (FGGE). Under this heading, the following papers were given :

- General objectives of GARP by Prof. Pierre Morel (France),
- Review of the GATE oceanographic programme by Prof. Gerold Siedler (Fed. Rep. of Germany),
- IOC Operational ocean services to GARP by Dr. W.N. English (Canada), and
- Buoy systems for the global experiment by Mr. J.E. Masterson (USA).

In addition to these lectures on planned programmes, presentations were made on recent or ongoing investigations :

Figure 17 - Some photographs taken on board the research vessel *Anton Bruun* during the International Indian Ocean Expedition:
1. Collecting water bottle samples.
2. Plankton net.
3. Dr. Panikkar (India)* examining preserved plankton.
4. Sorting specimens from trawl haul.
5. Removing mud sample from LaFond-Dietz bottom sampler.
(By courtesy of Dr. E.C. LaFond, Secretary, IAPSO)

* Chairman, IOC, 1964-1965.

- Main objectives and some results of the USSR National Expedition TROPEX-72 by Dr. M. R. Petrosianz (USSR), presented by Dr. V. M. Gruzinov (USSR), and
- Antilles-Guyana countercurrent by Prof. V. A. Kort (USSR) ;

and further, some information on the Co-operative Investigations in the Mediterranean (CIM) :

- General introductory remarks on CIM BY Dr. J. Joseph (Fed. Rep. of Germany) (International Co-ordinator for CIM)
- Physical oceanography of the Mediterranean by Prof. H. Lacombe (France),
- Geophysics of the Mediterranean by Prof. C. Morelli (Italy), and
- Pollution in the Mediterranean by Mr. O. Le Faucheux (France).

Finally, the Assembly was informed about the scientific results of the International Co-operative Investigations of the Tropical Atlantic (ICITA) and of the Guinean Trawling Survey (GTS) by the following Bruun Memorial Lectures :

- Introductory remarks as International Co-ordinator of ICITA by Dr. T.S. Austin (USA),
- The Equatorial Undercurrent in the Atlantic by Prof. Dr. K. Voigt (German Democr. Rep.),
- Plankton of the Tropical Atlantic by Prof. Dr. J. Krey (Fed. Rep. of Germany),
- Fishery resources of the tropical eastern-central Atlantic Ocean : Exploration, utilization and management since 1960 by Dr. F. Williams (USA), and
- Some results from the Eastern Atlantic Continental Margin Programme by Dr. K.O. Emery (USA).

These Bruun Memorial Lectures were published in *IOC Technical Series* No 11 (1975).

The practice of reporting on the scientific results of IOC's co-operative investigations was continued at the ninth session of the IOC Assembly in 1975 when the following Bruun Memorial Lectures were held :

- The Co-operative Study of the Kuroshio and adjacent regions (CSK) by Prof. K. Sugawara (Japan),
- Accuracy of the determination of nutrient elements in sea water by using CSK standard solutions by Dr. M. Ambe (Japan),
- Dynamics of the Kuroshio Current : Experimental and theoretical studies from south of Kyushu to the Izu-Ogasawara ridge by Prof. A.R. Robinson (USA), and
- Dynamics and thermohaline structure of the waters of the Kuroshio Current and adjacent regions by Prof. V.A. Kort (USSR).

These lectures were published in *IOC Technical Series* No 15 (1976). At the tenth session of the IOC Assembly in 1977, the Bruun Memorial Lectures were again devoted to forward-looking topics, namely to oceanography from space :

- Past, present and future capabilities of satellites relative to the needs of ocean science by Dr. J.R. Apel (USA),
- Remote sensing of the ocean in the USSR by Prof. Dr. B. Nelepo (Ukrainian SSR),
- The interrelation between electromagnetic waves and ocean waves by Dr. Werner Alpers (Fed. Rep. of Germany), and
- Determination of ocean tides from space by Dr. D. E. Cartwright (UK).

These lectures were published in the *IOC Technical Series* No 19 (1979)

In addition to these lectures at IOC Assembly sessions, it should be mentioned that some scientific contributions were also made at sessions of the IOC Executive Council. The relevant topics may be shortly summarized as follows.

At its fourth session in Ottawa, Canada in 1974, the Executive Council listened to 12 reports on the work going on under the umbrella of the International Decade of Ocean Exploration (IDOE). Information was provided on the Controlled Ecosystem Pollution Experiment (CEPEX), as well as on geophysical investigations around South America and the French-American Mid-Ocean Underwater Studies (FAMOUS). Further, the Council was informed about the results of the Joint North Sea Wave Project – JONSWAP'73, of the Overflow'73 investigations on the water exchange across the Greenland-Scotland Ridge, and on the Joint Air-Sea Interaction Project (JASIN). Finally, the Executive Committee' s interest was drawn to experiments with buoys tracked by satellite in Australian waters as well as to the variability of the sea surface temperature and salinity fields of the South-west Pacific Ocean.

At the fifth session of the IOC Executive Council in Venice, Italy in 1975, the scientific contributions dealt with the services of IOC, evaluating the importance to marine science of IGOSS, IODE and the Tsunami Warning System in the Pacific.

Finally, at the seventh session of the IOC Executive Council in Bergen, Norway in 1976, the lectures were concentrated on the scientific aspects of marine pollution, particularly monitoring.

Apart from such valuable information on and evaluation of the main research areas in oceanography, the need was felt in IOC for a proper assessment of the scientific content and the scientific priorities of IOC's own programmes. Although IOC, during all the years of its existence, has had the privilege of obtaining pertinent and helpful scientific advice from its advisory bodies, the Commission recognized that the desired analysis of the scientific content of IOC's programmes can best be provided by a body composed of a limited number of eminent marine scientists who, in their capacity as governmental experts, are fully aware of and familiar with the possibilities, necessities, peculiarities and limitations characterizing the work of an intergovernmental body like IOC.

In order to close this gap, the IOC, at the ninth session of its Assembly in 1975, decided to establish a *Scientific Advisory Board* and charged it with the following tasks (Annex to resolution IX-19) :

(a) to review the scientific content of the ongoing work of IOC subsidiary bodies, research programmes of LEPOR, workshop reports and proposals for new research programmes by Member States and regional groups, or any other matter referred to it by the Commission ;

(b) to report to IOC on the scientific strengths and weaknesses of all IOC activities, pointing out duplications of effort, gaps in ongoing projects and new opportunities for productive scientific programmes ;

(c) to make recommendations to the IOC on suggested priorities in scientific programmes, needs for publications and atlases and on other matters needed to strengthen the scientific aspects of the work of IOC.

The Scientific Advisory Board was established on an experimental basis for a period of two years, until 1977, when further assessment of its value was scheduled to take place. At the tenth session of the IOC Assembly in 1977, although the Board submitted a comprehensive report containing constructive criticism and useful recommendations on most of IOC's activities, opinions on its performance were divided. Therefore, the Board was asked to continue its work with an increased membership until the eleventh session of the Assembly when a final decision could be made on the need for an overall scientific subsidiary body for the Commission.

CHAPTER V

Future

16 Outlook on the future role and functions of the Commission

The review of IOC's history, functions and achievements, given in the preceding sections, has clearly shown that the role of IOC essentially depends on two kinds of influencing forces, namely :

- *Internal forces*, or the will expressed by IOC's Member States about the role IOC should play in the development of ocean research and the executive power behind such will ;
- *external forces*, or the requirements put to IOC from outside, i.e. from other international organizations and the degree of freedom available for responding to them adequately.

There is a certain relationship between internal and external forces in so far as most IOC Member States are also members of the international organizations which may formulate tasks for IOC. Experience has, however, shown that such a connexion is generally not very close.

As to the *internal forces*, it must be taken into account that among IOC's Member States there is a strong majority of developing countries which need advice and assistance in the development of their coastal zones and further ocean areas the exploration and economic exploitation of which will most probably fall under their jurisdiction once the new ocean regime is created. Consequently, IOC will have to shape its scientific programme in such a way that it is composed of *attractive, resource-oriented projects* wherein a maximum number of Member States is interested and able to participate fully from the planning stage onward. Such projects will generally not be found and realized globally, but in certain limited ocean regions where concurrence with regard to research goals and ways of implementation can best be achieved. These various research projects, planned and carried out regionally, can be integrated into a *new global programme* consisting of *investigations on the marine environment and resources in the ocean* and covering the decade from 1981 to 1990, thus representing the next phase of LEPOR after the termination of its acceleration phase IDOE in 1980.

The necessary prerequisite for such fruitful and trustful co-operation among Member States is that *real partnership* can be developed between marine scientists representing countries sufficiently developed in oceanography and those coming from countries which presently lack the necessary scientific personnel and infrastructure. To establish such partnership through mutual assistance should be the *foremost aim of IOC*. It will require a massive effort in the field of training, education and technical aid over at least a decade. It is hoped that the Voluntary Assistance Programme (IOC/VAP) will develop into an effective mechanism for realizing such mutual assistance, if the related activities of the ICSPRO organizations are co-ordinated adequately with IOC's endeavours. Only when the marine science capacities of all IOC Member States have been raised to a healthy and competent level, will international ocean research flourish and lead to results that are beneficial to all mankind.

Such IOC programmes will certainly require a great amount of understanding on the side of the industrialized countries which should be willing to pay due regard to the interests and needs of the developing world when planning research projects, in order to obtain maximum participation of Member States in them. It is evident that a proper balance between the interests of developed and developing states would be the best solution. Therefore, two avenues should be used in the conduct of marine scientific research : one, governed by the principle of optimum participation and oriented towards practical goals of interest and providing benefit to a great number of Member States, while the other approach can be characterized by very sophisticated research projects which are perhaps not of immediate interest to developing countries but promise important scientific progress in special fields.

In addition to taking care of the requirements of ocean science and TEMA, IOC should continue to pay due attention to the aspects of *ocean services*. Some services, as e.g. the systems of international oceanographic data exchange and of tsunami warning, appear to be well advanced and progressing satisfactorily. The Integrated Global Ocean Station System, however, still is in its infant stage, implying scientific, technical and economic problems but also representing a considerable challenge to the Commission which is the only intergovernmental body able to undertake such a difficult task. IGOSS needs more interested users, ingenious technical innovations, adequate predictive models of the ocean and sufficient financial support in order to become an effective working programme.

For carrying out its functions, the *standing of IOC* within the UN family is an important factor. At present, many arguments are in favour of keeping IOC within the framework of Unesco. Whilst remaining under the general umbrella of Unesco, IOC should, however, obtain much more support and, at the same time, as much autonomy from its parent agency as it needs in order to fulfil its role in international oceanography properly.

What will be this role ? This is a matter where *external forces* come into play. Among these external forces the introduction of a new ocean regime appears to be the most important one. At the present stage, it is by no means clear what additional tasks will emerge for IOC if the Third UN Conference on the Law of the Sea (UNCLOS) ends up in a proper international convention. But, considering what IOC has been doing and has achieved so far in the field of ocean research and ocean services, it is hardly conceivable that no new requirements will devolve on the Commission therefrom. In a number of articles, the Informal Composite Negotiating Text (ICNT) of

UNCLOS, presently available, indicates tasks and responsibilities for «competent marine scientific organizations» in promoting and implementing marine scientific research, transfer of marine technology and measures for the protection and preservation of the marine environment. As far as composition, expertise and experience are concerned, there is only one intergovernmental organization, namely the IOC, which has the capability to deal with all aspects of marine scientific research and related technical aid on a global scale. Therefore, it appears to be both logical and economical to avoid any unnecessary duplication of effort and to give IOC the focal role in this respect which would really entrust full responsibility to that international body which will and can ensure true internationalization of marine research.

ANNEX I

List of IOC Member States (as at I March 1979)

Algeria
Argentina
Australia
Austria
The Bahamas
Belgium
Brazil
Bulgaria
Cameroon
Canada
Chile
China
Colombia
Congo
Costa Rica
Cuba
Cyprus
Denmark
Dominican Republic
Ecuador
Egypt, Arab Republic of
Ethiopia
Fiji
Finland
France
Gabon
German Democratic Republic
Germany, Federal Republic of
Ghana
Greece
Guatemala
Guyana
Haiti
Iceland
India
Indonesia
Iran
Iraq
Ireland
Israel
Italy
Ivory Coast
Jamaica
Japan
Jordan
Kenya
Korea, Democratic People' s Republic of
Korea, Republic of
Kuwait
Lebanon
Libyan Arab Jamahiriya
Madagascar
Malaysia
Malta
Mauritania, Islamic Republic of
Mauritius
Mexico
Monaco
Morocco
Netherlands
New Zealand
Nicaragua
Nigeria
Norway
Pakistan
Panama
Peru
Philippines
Poland
Portugal
Qatar
Romania
Saudia Arabia
Senegal
Seychelles, Republic of
Sierra Leone
Singapore
Somalia
South Africa, Republic of (suspended)
Spain
Sri Lanka
Sudan
Surinam
Sweden
Switzerland
Syrian Arab Republic
Tanzania, United Republic of
Thailand
Togo
Tonga
Trinidad and Tobago
Tunisia
Turkey
Ukrainian SSR
Union of Soviet Socialist Republics
United Arab Emirates
United Kingdom
United States of America
Uruguay
Venezuela
Viet Nam, Socialist Republic of
Western Samoa
Yugoslavia

Total : 103 Member States

ANNEX II

List of Acronyms and Abbreviations

ACMRR	Advisory Committee on Marine Resources Research (FAO)
ACOMR	Advisory Committee on Oceanic Meteorological Research (WMO)
ASFIS	Aquatic Sciences and Fisheries Information System (FAO/IOC/UN)
BATHY	Bathythermograph report
BIOMASS	Biological Investigation of Marine Antarctic Systems and Stocks
CCOP	Committee for Co-ordination of Joint Prospecting for Mineral Resources in Asian Offshore Areas (ESCAP)
CCOP/SOPAC	Committee for Co-ordination of Joint Prospecting for Mineral Resources in South Pacific Offshore Areas (ESCAP)
CECAF	Fishery Committee for the Eastern Central Atlantic (FAO)
CEPEX	Controlled Ecosystem Pollution Experiment
CICAR	Co-operative Investigations of the Caribbean and Adjacent Regions (IOC)
CIG	International Geophysical Committee (ICSU)
CIM	Co-operative Investigations in the Mediterranean (IOC/FAO(GFCM)/ICSEM)
CINECA	Co-operative Investigations of the Northern Part of the Eastern Central Atlantic (ICES/IOC/FAO (CECAF))
CMG	Commission for Marine Geology (IUGS)
CSK	Co-operative Study of the Kuroshio and adjacent regions (IOC)
DNA	Designated National Agency (IOC/IODE)
DNP	Declared National Programme (IOC)
EC	Executive Council (IOC)
EC	Executive Committee (WMO)
ECOR	Engineering Committee on Oceanic Resources
ECOSOC	Economic and Social Council (UN)
EL NINO	Meteorological and oceanographic phenomenon off the west coast of South America
ESCAP	Economic and Social Commission for Asia and the Pacific
FAMOUS	French-American Mid-Ocean Underwater Studies
FAO	Food and Agriculture Organization of the United Nations
FGGE	First GARP Global Experiment (WMO/ICSU)
GARP	Global Atmospheric Research Programme (WMO/ICSU)
GATE	GARP Atlantic Tropical Experiment
GEBCO	General Bathymetric Chart of the Oceans (IOC/IHO)
GELTSPAP	Group of Experts on Long-term Scientific Policy and Planning (IOC)
GESAMP	Group of Experts on the Scientific Aspects of Marine Pollution (IMCO/FAO/Unesco/WMO/WHO/IAEA/UN/UNEP)
GFCM	General Fisheries Council for the Mediderranean (FAO)
GIPME	Global Investigation of Pollution in the Marine Environment (IOC)
GTC	Group for Technical Co-ordination (CIM)
GTS	Global Telecommunications System (WMO)
GTS	Guinean Trawling Survey
IAEA	International Atomic Energy Agency
IAPO	International Association of Physical Oceanography (IUGG)
IAPSO	International Association for the Physical Sciences of the Ocean (IUGG)
IBCM	International Bathymetric Chart of the Mediterranean
ICES	International Council for the Exploration of the Sea
ICG	International Co-ordination Group (IOC)
ICITA	International Co-operative Investigations of the Tropical Atlantic (IOC)
ICNAF	International Commission for the North-west Atlantic Fisheries
ICNT	Informal Composite Negotiating Text (UNCLOS)
ICSEM	International Commission for the Scientific Exploration of the Mediterranean Sea
ICSPRO	Inter-secretariat Committee on Scientific Programmes Relating to Oceanography (UN/FAO/Unesco/WMO/IMCO)
ICSU	International Council of Scientific Unions
IDOE	International Decade of Ocean Exploration
IDPSS	IGOSS Data Processing and Services System
IFRB	International Frequency Registration Board (ITU)
IGOSS	Integrated Global Ocean Station System (IOC/WMO)
IGY	International Geophysical Year (ICSU)
IHB	International Hydrographic Bureau (IHO)

IHO	International Hydrographic Organization
IIOE	International Indian Ocean Expedition (IOC)
IMCO	Inter-Governmental Maritime Consultative Organization
IOC	Intergovernmental Oceanographic Commission
IOCARIBE	IOC Association for the Caribbean and adjacent regions
IOC-VAP	IOC Voluntary Assistance Programme
IODE	International Oceanographic Data Exchange (IOC)
IOS	IGOSS Observing System
IPLAN	Joint IOC/WMO Planning Group for IGOSS
IRES	Group of Experts on Oceanographic Research as it relates to IGOSS (IOC)
ITECH	Joint IOC/WMO Group of Experts on IGOSS Technical Systems Design and Development and Service Requirements
ITEL	Joint WMO/IOC Group of Experts on Telecommunication (IGOSS)
ITIC	International Tsunami Information Center (IOC)
ITSU	IOC Tsunami Warning System in the Pacific
ITU	International Telecommunication Union
IUGG	International Union of Geodesy and Geophysics (ICSU)
IUGS	International Union of Geological Sciences (ICSU)
JASIN	Joint Air-Sea Interaction Project
JONSWAP	Joint North Sea Wave Project
LEPOR	Long-term and Expanded Programme of Oceanic exploration and Research (IOC)
MAOA	Executive Committee Panel on Meteorological Aspects of Ocean Affairs (WMO)
MAP	Mediterranean Action Plan (UNEP)
MAPMOPP	Marine Pollution (Petroleum) Monitoring Pilot Project (IGOSS)
MEDI	Marine Environmental Data Information Referral System (IOC)
NMC	National Meteorological Centre (WMO)
NODC	National Oceanographic Data Centre
ODAS	Ocean Data Acquisition Systems, Aids and Devices
RDC	Regional Data Centre
RNODC	Responsible National Oceanographic Data Centre
ROMBI	Results of Marine Biological Investigations (data format)
ROSCOP	Report of Observations/Samples Collected by Oceanographic Programmes (data format)
SAB	Scientific Advisory Board (IOC)
SCAR	Scientific Committee on Antarctic Research (ICSU)
SCOR	Scientific Committee on Oceanic Research (ICSU)
SEATAR	CCOP/IOC Joint Working Group on IODE Studies on East Asia Tectonics and Resources
SOC	International Co-ordination Group for the Southern Oceans
SOP	Special Observing Period (FGGE/GARP)
SOPAC	See : CCOP/SOPAC
TEMA	Training, Education and Mutual Assistance in the marine sciences (IOC)
TESAC	Temperature-Salinity-Currents (designator for radio message)
TROPEX	Tropical Experiment (USSR)
UN	United Nations
UNCLOS	(Third) UN Conference on the Law of the Sea
UNDP	UN Development Programme
UNEP	UN Environment Programme
Unesco	UN Educational, Scientific and Cultural Organization
VAP	Voluntary Assistance Programme, see : IOC-VAP
WARC	World Administrative Radio Conference for Maritime Mobile Telecommunications (ITU)
WDC	World Data Centre
WESTPAC	IOC Working Group for the Western Pacific
WHO	World Health Organization
WMARC	World Maritime Administrative Radio Conference (ITU)
WMO	World Meteorological Organization
WWW	World Weather Watch (WMO)

ANNEX III

LIST OF PUBLICATIONS RELATED TO IOC ACTIVITIES*

(a) IOC Technical Series

No. 1+ Manual on International Oceanographic Data Exchange. 1965. 28 p. (superseded)
No. 2+ Intergovernmental Oceanographic Commission (Five years of work). 1966. 39 p.
No. 3 Radio communication requirements for oceanography. 1967. 19 p.
No. 4+ Manual on International Oceanographic Data Exchange (Second edition). 1967. 49 p. (superseded by IOC Manuals and Guides No. 9)
No. 5+ Legal problems associated with Ocean Data Acquisition Systems (ODAS). 1969. 40 p.
No. 6 Perspectives in oceanography, 1968. 1969. 90 p.
No. 7 Comprehensive outline of the scope of the long-term and expanded programme of oceanic exploration and research. 1970. 82 p.
No. 8+ IGOSS (Integrated Global Ocean Station System), General plan and implementation programme for Phase I. 1971. 27 p. (superseded by No. 16)
No. 9+ Manual on International Oceanographic Data Exchange (Third Edition) 1973. 63 p. (superseded by IOC Manuals and Guides No. 9).
No. 10 Bruun memorial lectures (presented at the seventh session of the IOC, 1971). 1972. 43 p.
No. 11 Bruun memorial lectures, 1973 (presented at the eighth session of the IOC Assembly). 1975. 63 p.
No. 12 Oceanographic products and methods of analysis and prediction. 1977. 172 p. (English only)
No. 13 The International Decade of Ocean Exploration (IDOE) 1971-1980. 1974. 87 p.
No. 14 A comprehensive plan for the Global Investigation of Pollution in the Marine Environment and baseline study guidelines. 1976. 42 p.
No. 15 Bruun memorial lectures, 1975 (presented at the ninth session of the IOC Assembly). 1976. 59 p.
No. 16 Integrated Global Ocean Station System (IGOSS) General Plan and Implementation Programme 1977-1982. 37 p.
No 17 Oceanographic components of the Global Atmospheric Research Programme (GARP). 1977. 35 p.
No. 18 Global marine pollution : an overview. 1977. 96 p.
No. 19 Bruun memorial lectures, 1977 (presented at the tenth session of the IOC Assembly). 1979.
No. 20 History, functions and achievements of the Intergovermental Oceanographic Commission 1961-1977. 1979.
n/n Annotated bibliography of textbooks and reference materials in marine sciences. 1975. 109 p. (quadrilingual).

(b) IOC Manuals and Guides

No. 1 Manual on IGOSS data archiving and exchange. 1974. 69 p.
No. 2 International catalogue of ocean data stations. 1975. 83 p.
No. 2 International catalogue of ocean data stations - amendment No. 1. 1976. 100 p. (English only)
No. 3 Guide to operational procedures for the collection and exchange of oceanographic data (BATHY AND TESAC). 1975. 39 p.
No. 3 Guide to operational procedures for the collection and exchange of oceanographic data (BATHY and TESAC) - Amendment 2. 1978. 19 p. (incorporates Amendment 1)
No. 4 Guide to oceanographic and marine meteorological instruments and observing practices. 1975. 62 p. (English only).

No. 5 Guide to establishing a National Oceanographic Data Centre. 1975. 51 p.
No. 6 Wave reporting procedures for tide observers in the Tsunami Warning System. 1975. 32 p. (English and Spanish only).
No. 7 Guide to operational procedures for the IGOSS pilot project on marine pollution (petroleum) monitoring. 1976. 50 p.
No. 7 Suppl. Manual for monitoring of oil and petroleum hydrocarbons in marine waters and on beaches. 1977. 20 p. (English and French only).
No. 8+ Marine environmental data information referral catalogue (MEDI Pilot Catalogue). 1976. 146 p. (English only).
No. 9 Manual on International Oceanographic Data Exchange (fourth edition). 1976. 81 p.

---

* These publications are available in English, French, Spanish and Russian, except where specified.
+ These publications are out of stock.

(c) IOC Workshop Series

The Scientific Workshops of the Intergovernmental Oceanographic Commission are usually jointly sponsored with other intergovernmental or non-governmental bodies. In each case, by mutual agreement, one of the sponsoring bodies assumes responsibility for publication of the final report.

No.	*Title*	*Publishing Body*	*Languages*
1	Metallogenesis, Hydrocarbons and Tectonic Patterns in Eastern Asia (Report of an IDOE Workshop) ; Bangkok, Thailand, 24-29 September 1973. CCOP-IOC, 1974.	Office of the Project Manager UNDP/CCOP, c/o ESCAP Sala Santitham Bangkok 2, Thailand	English
2	Ichthyoplankton, Report of the CICAR Ichthyoplankton Workshop, Mexico City, 16-27 July 1974 (Unesco Technical Papers in Marine Sciences, No. 20).	Division of Marine Sciences, Unesco Place de Fontenoy 75700 Paris, France	English Spanish
3	Report of the IOC/FAO/(GFCM)/ICSEM International Workshop on Marine Pollution in the Mediterranean, Monte Carlo, 9-14 September 1974.	IOC, Unesco Place de Fontenoy 75700 Paris, France	English French Spanish
4	Workshop on the Phenomenon known as «El Niño», Guayaquil, Ecuador, 4-12 December 1974.	FAO Via delle Terme di Caracalla 00100 Rome, Italy	English Spanish
5	IDOE International Workshop on Marine Geology and Geophysics of the Caribbean Region and its Resources, Kingston, Jamaica, 17-22 February 1975.	IOC, Unesco Place de Fontenoy 75700 Paris, France	English Spanish
6	Report of the CCOP/SOPAC-IOC IDOE International Workshop on Geology, Mineral Resources and Geophysics of the South Pacific, Suva, Fiji, 1-6 September 1975.	IOC, Unesco Place de Fontenoy 75700 Paris, France	English
7	Report of the Scientific Workshop to initiate planning for a co-operative investigation in the North and Central Western Indian Ocean, organized within the IDOE under the sponsorship of IOC/FAO(IOFC)Unesco/EAC, Nairobi, Kenya, 25 March-2 April 1976.	IOC, Unesco Place de Fontenoy 75700 Paris, France	Full text (English only) Extract and Recommendations : French, Spanish, Russian
8	Report of the IOC/FAO (IPFC)/UNEP International Workshop on Marine Pollution in East Asian Waters, Penang, 7-13 April 1976.	IOC, Unesco Place de Fontenoy 75700 Paris, France	English

9	Report of the Second International Workshop on Marine Geoscience IOC/CMG/SCOR, Mauritius, 9-13 August 1976.	IOC, Unesco Place de Fontenoy 75700 Paris, France	English French Russian Spanish
10	Report of the Second IOC/WMO Workshop on Marine Pollution (Petroleum) Monitoring, Monaco, 14-18 June 1976.	IOC, Unesco Place de Fontenoy 75700 Paris, France	English French Russian Spanish
11	Report of the IOC/FAO/UNEP International Workshop on Marine Pollution in the Caribbean and Adjacent Regions, Port of Spain, Trinidad and Tobago, 13-17 December 1976.	IOC, Unesco Place de Fontenoy 75700 Paris, France	English Spanish
11 Suppl.	Collected Contributions of invited lecturers and authors to the IOC/FAO/UNEP International Workshop on Marine Pollution in the Caribbean and Adjacent Regions, Port-of-Spain, Trinidad and Tobago, 13-17 December 1976.	IOC, Unesco Place de Fontenoy 75700 Paris, France	English Spanish
12	Report of the IOCARIBE interdisciplinary workshop on scientific programmes in support of fisheries projects, Fort-de-France, Martinique, 28 November-2 December 1977, 37 p.	IOC, Unesco Place de Fontenoy 75700 Paris, France	English French Spanish
13	Report of the IOCARIBE Workshop on Environmental Geology of the Caribbean Coastal Area, Port-of-Spain, Trinidad and Tobago, 16-18 January, 1978. 30 p.	IOC, Unesco Place de Fontenoy 75700 Paris, France	English Spanish
14	Report of the IOC/FAO/WHO/UNEP International Workshop on Marine Pollution in the Gulf of Guinea and Adjacent Areas, Abidjan, Ivory Coast, 2-9 May 1978, 42 p.	IOC, Unesco Place de Fontenoy 75700 Paris, France	English French
15	Report of the CPPS/FAO/IOC/UNEP International Workshop on Marine Pollution in the Southeast Pacific, Santiago de Chile, 6-10 November 1978, 33 p.	IOC, Unesco Place de Fontenoy 75700 Paris, France	English Spanish
16	Workshop on the Western Pacific, Tokyo, 19-20 February 1979, 16 p.	IOC, Unesco Place de Fontenoy 75700 Paris, France	English French Russian
17	Report of the Joint IOC/WMO Workshop on Oceanographic Products and the IGOSS Data Processing and Services System (IDPSS), Moscow, 9-11 April 1979, 41 p.	IOC, Unesco Place de Fontenoy 75700 Paris, France	English

18	Syllabus for training marine technicians, Report of an IOC/Unesco workshop held in Miami, Florida, 22-26 May 1978. 50 p. (Unesco Reports in Marine Science, No. 4).	Division of Marine Sciences, Unesco Place de Fontenoy 75700 Paris, France	English French Russian Spanish
19	Marine science syllabus for secondary schools, Report of an IOC workshop held at United World College of the Atlantic, United Kingdom, 5-9 June 1978. 45 p. (Unesco Reports in Marine Science, No. 5).	Division of Marine Sciences, Unesco Place de Fontenoy 75700 Paris, France	English French Russian Spanish
20	Report and Recommendations of the Second CCOP-IOC Workshop on IDOE Studies of East Asia Tectonics and Resources, Bandung, Indonesia, 17-21 October 1978. 48 p.	CCOP c/o ESCAP United Nations Building Bangkok 2, Thailand	English

(d) Brochures

Intergovernmental Oceanographic Commission (Arabic, Chinese, English, French, Russian, Spanish).
IDOE 1971-1980 International Decade of Ocean Exploration.
Tsunami warning system in the Pacific (Chinese, English, French, Russian, Spanish).
Guide to international marine environmental data services.
General bathymetric chart of the oceans.
Aquatic Sciences and Fisheries Information System (Arabic, English, French, Russian, Spanish).
Pollution in the marine environment (Arabic, English, French, Russian, Spanish).

(e) Ad hoc publications

International Indian Ocean Expedition (IIOE)

International Indian Ocean Expedition, Collected Reprints, Volumes I to VIII, and Index. Various languages. Unesco.

The Biology of the Indian Ocean, Intenational Indian Ocean Expedition Symposium, Kiel, 31 March-6 April 1971. Edited by B. Zeitschel. Springer Verlag, Berlin, 1971 (Ecological Studies, vol. 3), English only.

Oceanographic Atlas of the International Indian Ocean Expedition, by K. Wyrtki, U.S. National Sciences Foundation, Washington, 1971, English only.

Meteorological Atlas of the International Indian Ocean Expedition :
Vol. I — The Surface Climate of 1963 and 1964, by C.S. Ramage, F.R. Miller and C. Jefferies
Vol. II — Upper Air, by C.S. Ramage and C.V.R. Raman.
A joint U.S. National Science Foundation and Indian Meteorological Department project. U.S. Government Printing Office, Washington, DC, 1972, English only.

International Indian Ocean Expedition : Handbook to the International Zooplankton Collections (curated and processed at the Indian Ocean Biological Centre). Vol III — Proceedings of the Workshop on Plankton Methods. National Institute of Oceanography, Cochin, India, 1972, English only.

The International Indian Ocean Expedition (1960-65) Oceanographic and Meteorological Monographs, University Press of Hawaii, Honolulu, Hawaii :

- Oceanographic Monographs
  1. The Monsoon Regime of the Currents in the Indian Ocean
- Meteorological Monographs
  1. Structure of an Arabian Sea Summer Monsoon System
  2. Average Cloudiness in the Tropics from Satellite Observations
  3. Meteorological Data Catalogue
  4. Tropical Indian Ocean Clouds
  5. An Investigation of Heat Exchange
  6. Profiles of Wind, Temperature and Humidity over the Arabian Sea
  7. Computations of Surface Layer Air Parcel Trajectories and Weather in the Oceanic Tropics (last vol. issued in 1975).

International Indian Ocean Expedition Geological/Geophysical Atlas, Academy of Sciences of the USSR. 1975. 151 p. Composite English/Russian.

Phytoplankton Production Atlas of the International Indian Ocean Expedition Universitätsdruckerei Kiel. 1976. 70 p. English only.

Indian Ocean : Collected data on primary production, phytoplankton pigments, and some related factors (This volume includes data obtained during the International Indian Ocean Expedition co-ordinated by the IOC (1959-1965)). Compiled by B. Babenerd and J. Krey. Universitätsdruckerei Kiel, Federal Republic of Germany. 1974. 521 p.

Co-operative Study of the Kuroshio and adjacent regions (CSK)

CSK Atlases (7 volumes issued up to 1977), Japan Oceanographic Data Center, Tokyo, Japan.

Symposium on the Co-operative Study of the Kuroshio and adjacent regions (CSK) organized through the joint efforts of Unesco, FAO and the East-West Center, Honolulu, Hawaii, 29 April-2 May 1968. Reports and abstracts on papers. FAO Fisheries Report No. 63, 1968, English only.

The Kuroshio — A symposium on the Japan Current. Edited by J.C. Marr. (Includes 58 of the papers presented to the first Symposium on the Co-operative Study of the Kuroshio and adjacent regions, May 1968, Honolulu, Hawaii). East-West Center-Press, Honolulu, Hawaii, 1970, English only.

Kuroshio II - Proceedings of the Second Symposium on the results of the Co-operative Study of the Kuroshio and adjacent regions. Edited by K. Sugawara, Saikon Publ. Co. Ltd., Tokyo, 1972, English only.

Kuroshio III - Proceedings of the Third CSK Symposium, Bangkok, Thailand, 26-29 May 1973. Mongkol Karnpim Press and Publ., Bangkok, 1974, English only.

International Co-operative Investigations of the Tropical Atlantic (ICITA)

Proceedings of the Symposium on Oceanography and Fisheries Resources of the Tropical Atlantic (Results of the ICITA and GTS). Organized through the joint efforts of Unesco, FAO and OAU. Abidjan, Ivory Coast, 20-28 October 1966. Review papers and contributions, Unesco, 1969. Composite English/French.

Proceedings of the Symposium on the oceanography and fisheries resources of the Tropical Atlantic (Results of ICITA and GTS). Organized through the joint efforts of Unesco, FAO and OAU, Abidjan, Ivory Coast, 20-28 October 1966. Reports and abstracts of papers. FAO Fisheries Report No. 51, 1967 in English and French.

ICITA Atlas Volume 1 Physical Oceanography, Unesco. 1973 (quadrilingual)

ICITA Atlas Volume 2 Chemical and Biological Oceanography, Unesco. 1976 (quadrilingual)

Co-operative Investigations of the Caribbean and Adjacent Regions (CICAR)

Symposium on investigations and resources of the Caribbean Sea and adjacent regions. Preparatory to Co-operative Investigations in the Caribbean and Adjacent Regions (CICAR). Organized through the joint efforts of FAO and Unesco. Willemstad, Curaçao, Netherlands Antilles, 18-26 November 1968. Report and abstracts of papers. FAO Fisheries Report No. 71.1, 1969 (English and Spanish).

Symposium on investigations and resources of the Caribbean Sea and adjacent regions. Preparatory to Co-operative Investigations in the Caribbean and Adjacent Regions (CICAR). Organized jointly by FAO and Unesco. Willemstad, Curaçao, Netherlands Antilles, 18-26 November 1968. Papers on fishery resources. FAO Fisheries Report No. 71.2 (composite English/Spanish).

Symposium on Investigations and Resources of the Caribbean Sea and Adjacent Regions. Preparatory to CICAR, organized jointly by Unesco and FAO, Willemstad, Curaçao, Netherlands Antilles, 18-26 November 1968. Papers on Physical and Chemical Oceanography, Marine Geology and Geophysics and Marine Biology. Unesco, 1971 (composite English/Spanish).

Progress in marine research in the Caribbean and adjacent regions CICAR-II Symposium. Proceedings : Volume I - Papers on physical and chemical oceanography, marine geology and geophysics) ; Volume II - Papers on fisheries, aquaculture and marine biology, FAO Fisheries Report No. 200 (composite English/Spanish).

El Niño

Workshop on the Phenomenon known as «El Niño», Guayaquil, Ecuador, 4-12 December 1974, Proceedings, Spanish language papers presented, FAO Fisheries Report No. 185, 1976 (Spanish only).

Workshop on the Phenomenon known al «El Niño», Guayaquil, Ecuador, 4-12 December 1974, Proceedings, English language papers presented, Unesco (in press) (English only).

Miscellaneous

Safety Provisions of Ocean Data Acquisition Systems, Aids and Devices (ODAS), Unesco/IMCO, published by IMCO, London, 1972.

The Health of the Oceans, by E.D. Goldberg, Unesco 1976, English, Italian (French and Spanish in press).

Marine Affairs : Register of Courses and Training Programmes (UN No. ST/ESA/54), 1976 (English only).

Oceanic Water Balance (WMO - No. 442), 1976 (English only).

Marine Pollution Monitoring (Petroleum) — Proceedings of a symposium sponsored by IOC/WMO/US Department of Commerce, 1974 (NBS Special Publication 409) (English only).

Manual of Interpretation of Orbital Remote Sensing Satellite Photography and Imagery for Coastal and Offshore Environmental Features (including Lagoons, Estuaries and Bays). University of Munich, English only.

Annex IV

Preparation for the launching of a Bathythermograph aboard the Indian research ship KISTNA.

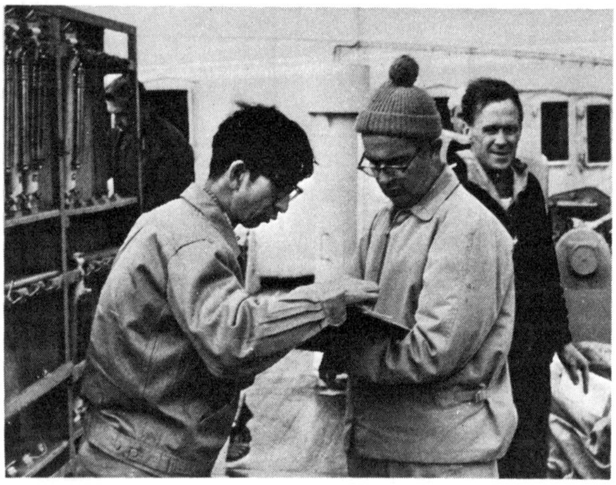

Marine scientists from Japan (Dr. Yoshiwaw Sugiyura), Australia (Dr. David Rochford) and USSR (Dr. Alex Bogojolensky) working together aboard the USSR research vessel VITYAZ.

Indian scientist sorts out samples of plankton, taken by the British research vessel DISCOVERY, in the Indian Ocean Biological Centre.

In order to inform marine scientists on the effects of marine pollution, Prof. Johannes Krey, a specialist in zooplankton, shows what has happened to oxygen levels in the Baltic Sea over the years.

Trainees from nine Asian countries take part in practical work on board the research ship STRANGER of the United States.

Tens of thousands of jars on these shelves contain specimens of the flora and fauna of the China Sea (Institute of Oceanography, Nathrang, Viet Nam).

Marine laboratory on board the Brazilian research training vessel ALMIRANTE SALDANHA which was equipped by Unesco.

Weight stand of a piston-corer being prepared for lowering aboard the US research ship ARGO.

STD (salinity, temperature, depth) recorder being lowered in the Caribbean.

Monster-Buoy developed in the United States for recording meteorological and oceanographic parameters.